T0211913

Understanding Atrial Fibrillation:
The Signal Processing Contribution
Part II

SYNTHESIS LECTURES ON BIOMEDICAL ENGINEERING

Editor
John D. Enderle, *University of Connecticut*

Understanding Atrial Fibrillation: The Signal Processing Contribution, Part II
Luca Mainardi, Leif Sörnmo, and Sergio Cerutti
2008

Lung Sounds: An Advanced Signal Processing Perspective
Leontios J. Hadjileontiadis
2008

An Outline of Information Genetics
Gérard Battail
2008

Neural Interfacing: Forging the Human-Machine Connection
Thomas D. Coates, Jr.
2008

Quantitative Neurophysiology
Joseph V. Tranquillo
2008

Tremor: From Pathogenesis to Treatment
Giuliana Grimaldi and Mario Manto
2008

Introduction to Continuum Biomechanics
Kyriacos A. Athanasiou and Roman M. Natoli
2008

The Effects of Hypergravity and Microgravity on Biomedical Experiments
Thais Russomano, Gustavo Dalmarco, and Felipe Prehn Falcão
2008

Understanding Atrial Fibrillation: The Signal Processing Contribution, Part II

Luca Mainardi, Leif Sörnmo, and Sergio Cerutti

ISBN: 978-3-031-00504-6   paperback
ISBN: 978-3-031-01632-5   ebook

DOI 10.1007/978-3-031-01632-5

A Publication in the Springer series
SYNTHESIS LECTURES ON BIOMEDICAL ENGINEERING

Lectures #25
Series Editor: John D. Enderle, *University of Connecticut*

Series ISSN
Synthesis Lectures on Biomedical Engineering
Print 1930-0328   Electronic 1930-0336

Understanding Atrial Fibrillation: The Signal Processing Contribution Part II

Luca Mainardi
Department of Bioengineering
Politecnico di Milano, Italy

Leif Sörnmo
Department of Electrical Engineering
Lund University, Sweden

Sergio Cerutti
Department of Bioengineering
Politecnico di Milano, Italy

SYNTHESIS LECTURES ON BIOMEDICAL ENGINEERING #25

ABSTRACT

The book presents recent advances in signal processing techniques for modeling, analysis, and understanding of the heart's electrical activity during atrial fibrillation. This arrhythmia is the most commonly encountered in clinical practice and its complex and metamorphic nature represents a challenging problem for clinicians, engineers, and scientists. Research on atrial fibrillation has stimulated the development of a wide range of signal processing tools to better understand the mechanisms ruling its initiation, maintenance, and termination. This book provides undergraduate and graduate students, as well as researchers and practicing engineers, with an overview of techniques, including time domain techniques for atrial wave extraction, time–frequency analysis for exploring wave dynamics, and nonlinear techniques to characterize the ventricular response and the organization of atrial activity. The book includes an introductory chapter about atrial fibrillation and its mechanisms, treatment, and management. The successive chapters are dedicated to the analysis of atrial signals recorded on the body surface and to the quantification of ventricular response. The rest of the book explores techniques to characterize endo- and epicardial recordings and to model atrial conduction. Under the appearance of being a monothematic book on atrial fibrillation, the reader will not only recognize common problems of biomedical signal processing but also discover that analysis of atrial fibrillation is a unique challenge for developing and testing novel signal processing tools.

KEYWORDS

Atrial fibrillation, signal processing, signal modeling, atrial activity extraction, spectral analysis, time–frequency analysis, ventricular response characterization, heart rate variability, atrial organization indices, source modeling, volume conductor modeling, intracardiac AF detection, AF long-term monitoring, implantable cardiac devices pacemakers.

AUTHORS

Andreas Bollmann
Department of Electrophysiology, Leipzig University Heart Center, Leipzig, Germany

Andreu Climent
Electronic Engineering Department, Valencia University of Technology, Valencia, Spain

Valentina Corino
Department of Bioengineering, Politecnico di Milano, Milano, Italy

Luca Faes
Department of Physics, University of Trento, Trento, Italy

Jeff Gillberg
Medtronic Inc., Minneapolis, Minnesota, USA

Vincent Jacquemet
Department of Cardiology, University of Lausanne, Lausanne, Switzerland

Federico Lombardi
Division of Cardiology, San Paolo Hospital, University of Milan, Milan, Italy

Luca Mainardi
Department of Bioengineering, Politecnico di Milano, Milano, Italy

Rahul Mehra
Medtronic Inc., Minneapolis, Minnesota, USA

Adriaan van Oosterom
Department of Cardiology, University of Lausanne, Lausanne, Switzerland

Simona Petrutiu
Department of Electrical Engineering, Northwestern University, Chicago, USA

Flavia Ravelli
Department of Physics, University of Trento, Trento, Italy

José Joaquín Rieta
Biomedical Synergy, Valencia University of Technology, Gandia, Spain

Alan Sahakian
Department of Electrical Engineering, Northwestern University, Chicago, USA

Frida Sandberg
Department of Electrical Engineering, Lund University, Lund, Sweden

Shantanu Sarkar
Medtronic Inc., Minneapolis, Minnesota, USA

Leif Sörnmo
Department of Electrical Engineering, Lund University, Lund, Sweden

Martin Stridh
Department of Electrical Engineering, Lund University, Lund, Sweden

Steven Swiryn
Division of Cardiology, Northwestern University, Chicago, USA

Paul Ziegler
Medtronic Inc., Minneapolis, Minnesota, USA

Contents

Part I 1

Andreas Bollmann and Federico Lombardi

Simona Petrutiu, Alan Sahakian, and Steven Swiryn

Preface

Atrial fibrillation (AF) is a widely diffused arrhythmia which affects 2 million people in Europe and roughly 2.2 million in the U.S., reducing quality of life and increasing risk for stroke and death. Atrial fibrillation is the most common cause of hospitalization and its treatment is one of the most cogent issues in clinical arrhythmology. Considerable research effort is currently directed to this arrhythmia because the mechanisms causing its initiation, maintenance, and termination are not sufficiently well understood. In addition, as AF prevalence and incidence doubles with each decade beyond 50 years of age, the impact of this arrhythmia will be progressively larger in the near future due to the aging population.

The complex, heterogeneous, and metamorphic nature of this arrhythmia represent a challenging problem for clinicians, engineers, and scientists, stimulating the development of quantitative methods for the analysis of the electrical signals recorded during AF. These are actually considerably larger than those employed to analyze any other cardiac arrhythmias.

Nevertheless, ECG-based analysis of AF has, for a long time, been confined to a qualitative characterization of the fibrillatory waves. Only in the last decade has more detailed information contained in the fibrillatory waves emerged, leading to more detailed characterization of this arrhythmia. This progress is largely due to the advent of reliable signal processing procedures which are able to separate atrial and ventricular activities.

This book presents a comprehensive overview of signal processing methods employed for modeling, analysis, and management of the heart's electrical activity during AF. While some signal processing challenges are common to those encountered in the analysis of other bioelectrical signals, other challenges are unique to the analysis of AF. The topics are related to spectral analysis (traditional and time-varying), noise-reduction, separation of signal sources, and assessment of nonlinear dynamics. We are therefore confident that this book will be of interest to researchers which are familiar with biomedical signal processing and to students who wants to learn about the new applications and challenges of biomedical signal processing. In addition, our hope is that this book will serve as an inspiration to further research on AF—a research area which is still in need of much exploration.

CONTENTS OVERVIEW

Chapter 1 presents current clinical knowledge about human AF. Starting from electrophysiological mechanisms and determinants of AF, the chapter considers AF incidence and prevalence and AF classification schemes used in clinical practice. The clinical consequences of AF as well as various therapeutic options are also described.

Chapter 2 describes time domain methods designed to characterize the atrial fibrillatory waveforms in ECG. Time domain parameters such as amplitude and rate are considered as well as their reproducibility over time. The value of vectorcardiographic analysis of f-waves is discussed at length as is the choice of lead configurations which are particularly suitable for analysis of AF.

Chapter 3 addresses with the fundamental problem of how to extract the atrial fibrillatory activity from the ECG. Different approaches for atrial extraction are reviewed, ranging from basic average beat subtraction to principal and independent component analysis. The problem of how to evaluate and compare the performance of different algorithms is considered.

Chapter 4 deals with the estimation of AF frequency, i.e., the repetition rate of the f-waves. This parameter has been found valuable in various clinical applications. Different techniques for time–frequency analysis of the atrial signal are presented, designed to unveil temporal variations in AF frequency which are either spontaneous in nature or due to some kind of intervention.

Chapter 5 summarizes methods for analysis of the ventricular response during AF and description of AV conduction characteristics. Such analysis is based on the description of RR interval variability using various signal processing tools of increasing complexity: from simple statistical indices to more sophisticated metrics derived from nonlinear dynamics theory.

In Chapter 6, the focus is shifted from the body surface to endo- and epicardial recordings. The problem of assessing "atrial organization" is introduced and as this term assumes different meanings, particular attention is paid to the peculiar characteristics of organization captured by each method. These characteristics include repeatability/regularity of the atrial activations, correlation/synchronicity among electrograms recorded in different locations, and similarity of the atrial wave morphology.

Chapter 7 introduces a computer model suitable for the interpretation and analysis of fibrillatory ECGs. Starting from a biophysical model of atrial cell membrane during AF, the basics of the forward problem are considered and used to generate simulated ECG signals which mimic the ones observed on the thorax. The model based approach provides a link between body surface recordings and atrial activity and puts into evidence the information content of these signals.

Chapter 8 deals with the problem of detecting atrial tachycardia/atrial fibrillation in implantable device, being crucial for long-term monitoring in which high specificity is usually required to avoid over-treatment of nonsustained atrial arrhythmias. After having introduced the clinical requirements of such a monitoring device, the chapter presents the details of a detector for atrial tachyarrhythmias which is based on the analysis of RR interval irregularity.

ACKNOWLEDGEMENTS

This book was made possible because of the collaboration between scientists who have made contributions to atrial fibrillation research. We wish to express our sincere gratitude to all the contributors for generously sharing their expertise. Without their great enthusiasm and willingness to spend precious time on writing, this book would never have happened.

We are also grateful to Dr. Sih who provided material for replicating Fig. 6.1 in Chapter 6.

Luca Mainardi
Leif Sörnmo
Sergio Cerutti
December 2008

Part II

CHAPTER 5

Analysis of Ventricular Response during Atrial Fibrillation

Valentina Corino, Andreu Climent, Luca Mainardi, and Andreas Bollmann

5.1   INTRODUCTION

During atrial fibrillation (AF), the fibrillatory impulses continuously bombard and penetrate the atrioventricular (AV) node to varying degrees (concealed conduction), creating appreciable variability on the AV nodal refractoriness [1]. Since the AV node is the structure responsible for the conduction of atrial impulses to the ventricles, the strategy of rate control during AF deals with efforts to utilize and adjust the propagation properties of the node [2]. Characteristics of AV conduction have been investigated in different species and with different techniques [3]. However, AV conduction mechanisms are not completely understood and, in particular, the role of the AV node physiology and the rate and irregularity of the atrial fibrillatory waves are still unknown.

Virtually in every patient with AF, standard 12-lead surface ECG and/or Holter recordings are acquired, the main purposes being confirmation of arrhythmia presence and determination of ventricular rate. However, even if the irregular ventricular response is often described as chaotic and without any form of patterning, previous studies have shown that this process is not completely random [4]. The ventricular response during AF has shown preferential conduction and different degrees of short-term predictability [5]. Signal processing plays an essential role for deeper understanding of AF mechanisms and for quantifying AF patterns and properties [4].

The present chapter summarizes the state-of-art in ventricular response analysis during AF as characterized by a variety of methods. In particular, processing techniques are introduced with increasing complexity: RR interval histograms, Poincaré plots, time domain heart rate parameters, spectral analysis and $1/f$ power law behavior, and nonlinear analysis. The latter type of analysis includes embedding time series derived methods, entropy, and regularity.

5.2   RR INTERVAL HISTOGRAMS

AV conduction properties can be evaluated by constructing RR interval histograms from Holter recordings of patients with paroxysmal and persistent AF. Histograms during AF may evidence uni-, bi-, or multimodal RR histogram patterns [4, 2, 6, 7, 8, 9]. It has been accepted that a bimodal RR interval histogram during AF suggests the presence of dual AV nodal physiology and predicts better outcome of radiofrequency ablation of the posterior atrionodal input [6, 7, 8, 9, 10, 11]. Information on electrophysiological characteristics such as functional refractory period (FRP) can be obtained noninvasively.

5.2.1 HEART RATE STRATIFIED HISTOGRAMS

As the RR intervals span over a wide range, a special technique for displaying and analyzing RR interval histograms is the so-called heart rate stratified histogram (HRSH) analysis. The HRSH has been presented as noninvasive support for the identification of multiple intranodal pathways in patients with AF [6, 7, 12, 13, 14]. This analysis allows a detailed observation of the RR distribution at different average heart rate levels and is able to detect dual distributions of RR intervals that are not appreciable by visual inspection in the unstratified histogram. RR interval histograms are constructed on the basis of mean heart rate of the analyzed ECG segment that contains a fixed number of beats. Premature ventricular beats with aberrant conduction or beats with RR intervals longer than 1,500 ms, due to nodal escape beats, are excluded from histogram analysis. Whenever noise is identified or when a beat is not classified as normal, the preceding and following intervals are discarded from the analysis [13].

The RR interval series is divided into overlapping sequences of length L:

$$
\begin{aligned}
\mathbf{x}^L(n) &= \begin{bmatrix} x(n) & x(n+1) & \cdots & x(n+L-1) \end{bmatrix}^T, \\
\mathbf{x}^L(n+k) &= \begin{bmatrix} x(n+k) & x(n+k+1) & \cdots & x(n+k+L-1) \end{bmatrix}^T,
\end{aligned} \tag{5.1}
$$

where $x(n)$ denotes RR interval and k is the distance in beats between two consecutive sequences. If $k < L$, then an overlap is present.

According to its mean, each sequence $\mathbf{x}^L(n)$ is classified into a heart rate level of beats per minute (bpm). Obviously, the histogram shape of each heart rate level depends on L, k, and the range that defines the heart rate level itself.

The output of this method is one histogram per heart rate level. Each histogram can be analyzed by visual inspection in regard to the number of distinct populations of RR intervals. A definition of the presence of a bimodal histogram was first made by Olsson et al. [6]. A peak is defined when it occurs in two adjacent histograms. Sixty-four RR intervals are used to construct the sequences, and ranges of 10 bpm to classify histograms in heart rate level, i.e., 60–70, 70–80, and so on. The HRSH method is designed to emphasize the presence of two RR populations only, neglecting the possible existence of additional populations. From Figure 5.1, it can be seen that the peak corresponding to the longer RR interval population is dominant at lower heart rate levels. As the heart rate increases, the peak of the shortest RR interval population increases and becomes dominant. The heart rate level at which the maximum amplitude of the histogram changes from one peak to another is defined as a *peak dominant change* value and is used for further analysis.

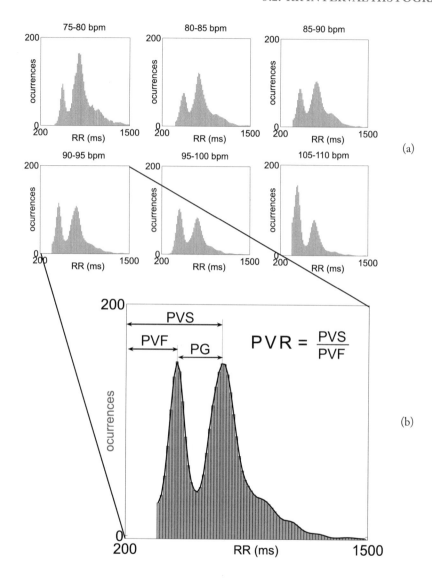

Figure 5.1: (a) Heart rate stratified histograms (HRSH) presenting two RR interval populations and the variation of the peak dominance with the heart rate level. (b) HRSH parameters evaluated for the peak dominant change (PDC) histogram. PVF denotes the peak value of the shortest RR interval population, PVS denotes the peak value of the longest RR interval population, PG denotes the peak gap which is the distance between the longest and shortest RR interval populations, and PVR denotes the peak value ratio which is the ratio between the longer and shorter RR interval populations. In this particular patient, PDC was 90–95 bpm, PVF = 480 ms, PVS = 720 ms, PVR = 1.5, and PG = 240 ms.

A smoothing procedure can be applied with the purpose of removing all local maxima except those belonging to the main RR populations. The three-term moving average recalculates each bin as:

$$
b_{j+1}(k) = \begin{cases} \frac{1}{3} \sum_{i=k-1}^{k+1} b_j(i), & k = 2, \ldots, K-1; \\ \\ b_j(k), & k = 1, K; \end{cases} \tag{5.2}
$$

where $b_j(k)$ is the kth bin of the histogram after j iterations of the filter, and K is the total number of bins in the histogram [13, 14]. Smoothing is applied until no local maxima are found within a distance of less than 50 ms. It is worth noting that this iterative procedure is equivalent to lowpass filtering with varying cut-off frequencies which depends on the number of iterations. After smoothing, the remaining peaks are considered to belong to different RR interval populations. The peak location of each population is estimated by fitting a second-degree polynomial to the highest bin and the four adjacent bins (two on each side). Finally, the highest and second highest peaks are classified as dominant and nondominant, respectively. The peaks corresponding to the longest and shortest RR interval populations are named *slow peak value* and *fast peak value*, respectively. The distance between the peak locations of the longest and shortest RR populations is the *peak gap*. The *peak value ratio* is the ratio between the slow and the fast peak value; see Figure 5.1.

5.2.2   APPLICATIONS

The minimal RR interval in chronic AF can be assumed to approximate the FRP of the AV node. Billete et al. described the 5th percentile RR interval (i.e., the lowest 5% RR intervals) during induction of AF to be equal to the FRP of the AV node when the FRP was short [15]. Khand et al. used the 5% of hourly histograms as a noninvasive marker for quantifying the dynamic interplay between sympathetic and parasympathetic inputs to the AV nodal compact region by means of the variations of its FRP [16]. On the other hand, the bimodal RR histogram has been thought to indicate dual pathways AV nodal during AF [2]. Although a bimodal histogram could be explained by the concealed conduction model in which AV node refractoriness differs between the proximal and distal region [17], Olsson et al. strongly suggested that the two RR populations correspond to conduction through two different atrionodal conduction routes [6].

Heart rate stratified histogram analysis has been applied to evaluate the effects of magnesium on AV nodal conduction [14]. In this study, magnesium at high concentrations caused a delay in both the shorter and longer RR intervals with a pronounced different effect on the two RR populations.

By means of HRSH, a bimodal RR interval histogram was found in 26 out of 32 individuals with persistent AF [7]. The main advantage of this technique over simple RR interval histogram analysis is that the identification of multiple RR populations is markedly improved. However, the bimodal RR interval histogram analysis is not always related to the presence of dual AV nodal pathway physiology, and a unimodal histogram may still be associated with dual AV nodal pathways [18]. Nevertheless, modifications of the AV node in patients with bimodal histograms were more effective in reducing the heart rate than in patients with unimodal ones [8]. These findings point to the possibility of predicting better specific treatments by means of HRSH analysis.

5.3   THE POINCARÉ PLOT

The RR interval histogram represents the distribution of single intervals, but gives no information about the sequence of intervals, i.e., the regularity of the ventricular rhythm. In the Poincaré plot, each RR interval is plotted versus the preceding one [19, 20, 21]; it should be noted that this type of plot is also referred to as the Lorenz plot, cf. Chapter 8. The pattern of such a plot can be inspected to distinguish AF from other supraventricular tachycardias such as atrial flutter where ventricular response is not as irregular as in AF. As shown in Figure 5.2(a), during sinus rhythm (SR), successive RR intervals are centered around the main diagonal forming an ellipsoid-like pattern, and each RR interval is strongly dependent on the preceding RR interval. Atrial flutter is characterized by organized atrial dysfunction and occurs when an abnormal conduction circuit develops inside the right atrium, with an activation rate of 250–300 bpm. The AV node conducts a periodic number of these activations which become visible as clusters in the Poincaré plot; see Figure 5.2(b) [22]. During AF, the irregularity of RR intervals results in a widely scattered distribution (Figure 5.2(c)) which is representative of disorganized atrial activity combined with atrioventricular conduction properties [4, 23, 22, 24, 25, 26, 27].

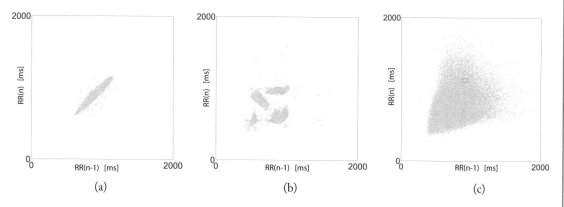

(a)   (b)   (c)

Figure 5.2: Poincaré plots of 24-h recordings of different patients during (a) sinus rhythm, (b) atrial flutter, and (c) atrial fibrillation.

It is possible to evaluate the lower envelope (LE) of the Poincaré plot as a measure of cycle length dependence on the FRP of the AV node [15]. Different techniques can be used to evaluate LE; following manual computation [20, 21, 28], linearly regression was proposed [25, 26, 27]. However, this method is dependent on the number of points used and is not robust against outliers. A new method based on the Hough transform was recently presented for automatic evaluation of the LE [29]; see Section 5.3.2.

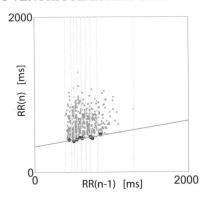

Figure 5.3: Poincaré plot of 512 successive RR intervals during AF. Vertical lines represent the separation between eight different bins, each one including 64 points. Black points are the minimal values of each bin. The lower envelope (solid line) is given by the regression line for the eight minimal values of each bin.

5.3.1   THE LINEAR REGRESSION LINE

Hayano et al. studied segments of 512 RR intervals every five minutes to analyze 24-h Holter recordings [25]. The horizontal axis, displaying $RR(n-1)$, was divided into 8 consecutive bins of 64 points; in each bin, the minimal value was determined. The LE was computed as the linear regression line fitting the eight minimal points. For episodes longer than 512 points, a larger number of points per bin or even more horizontal axis divisions could be used. However, this technique is not robust against outliers, e.g., due to artefacts or misdetected QRS complexes, as the regression line weighs all points equally.

5.3.2   THE HOUGH TRANSFORM

The Hough transform was proposed to detect straight lines in an image based on the representation of straight lines in the image space (x, y) using the slope–intercept equation $y = mx + c$ [30]. In the so-called Hough space, straight lines are characterized by the pair of parameters (m, c). Since slope and intercept are unbounded, an improvement of this method was proposed in [31] where the normal parameters (ρ, θ) were introduced as $\rho = x \cos \theta + y \sin \theta$. This improved method was used to calculate the LE of the Poincaré plot [29]. For a given $RR(n-1) = RR_0$, there are usually more corresponding $RR(n)$ values, thus the minimum RR_{min} of $RR(n)$ is selected for RR intervals not exceeding 2,000 ms (Figure 5.4(b)). Each pair of coordinates is transformed into the Hough space by

$$\rho(\theta) = RR_0 \cos \theta + RR_{min} \sin \theta, \quad -\frac{\pi}{2} \leq \theta \leq \frac{\pi}{2}. \tag{5.3}$$

The maximum point in the Hough space, indicated by the coordinates ρ_{max} and θ_{max}, is converted to the original space as:

$$(x_0, y_0) = \left(\frac{\rho_{max}}{\cos \theta_{max}}, \frac{\rho_{max}}{\sin \theta_{max}} \right). \tag{5.4}$$

The LE line is characterized by the slope m and the intercept c,

$$(m, c) = \left(-\frac{y_0 - 1}{x_0 - 1}, y_0 \right). \tag{5.5}$$

Principal advantages of this method are that the LE can be evaluated independently of the number of RR intervals and that misdetected QRS complexes, representing distant points in the Hough space, are automatically excluded.

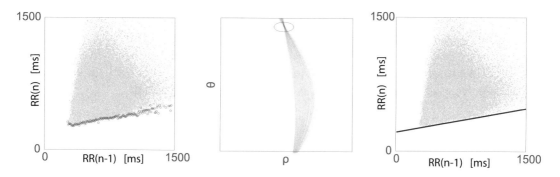

Figure 5.4: Evaluation of the lower envelope in a Poincaré plot using the Hough transform. (a) For each $RR(n-1)$ the corresponding minimum $RR(n)$ is found thus originating the pairs (RR_0, RR_{min}) (black circles). (b) Each pair (RR_0, RR_{min}) is transformed in the Hough space and the maximum is found (center of the circled point). (c) The lower envelope (black line) computed by the Hough transform is superimposed on the Poincaré plot. The parameters of the line that best fits this image are measured as the maximum in the Hough space (center of the circled point in (b)). Converting the coordinates of the darkest point in the Hough space ($\rho = -239$, $\theta = -1.46$ rad) into rectangular space, the slope of the LE is $m = 0.1107$ and its intersection with $RR(n-1) = 0$ is $c = 241$ ms.

5.3.3   APPLICATIONS

The LE, the slope m, and the 1-s intercept point have been used to characterize the FRP and the rate dependence of AV node conduction [24, 25, 26, 29]. In addition, the degree of scatter of the Poincaré plot, calculated as the root mean square difference of each RR interval and the LE, was presented as a measure of concealed conduction in the AV node [25].

By applying Poincaré plot analysis to 24-h Holter recordings of 48 patients with chronic AF, it was suggested that both AV node refractoriness and the degree of concealed AV conduction during AF may exhibit a circadian rhythm, but also that circadian rhythms may be attenuated in patients with heart failure [25]. These findings point to the possibility of obtaining information concerning altered autonomic control of the RR intervals in patients with AF (and even for heart failure or other disease) with this simple technique.

5.4   CLUSTERS IN THE POINCARÉ PLOT

Patients with AF and bimodal histograms would exhibit double sector shapes or superimposed scattergrams in Poincaré plots [27]. These shapes have been evaluated by means of Poincaré plots combined with histographic information [22].

5.4.1   DOUBLE SECTORS IN THE POINCARÉ PLOT

Oka et al. interpreted double sector shapes as the representation of the dual AV nodal pathway theory, and strongly suggested that Poincaré plots with double sectors hold information of the FRP of each of the two conduction routes [27]. The vertices of the two sectors were determined manually. In Figure 5.5, the leftmost vertex was considered to represent the FRP of the slower pathway, whereas the rightmost vertex was considered to represent the FRP of the faster pathway.

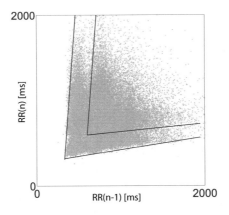

Figure 5.5: Determination of two vertices of the Poincaré plot with double sectors. Two vertices were studied as if they hold information on the FRP of each of the two conduction routes.

5.4.2   THE HISTOGRAPHIC POINCARÉ PLOT

By adding the number of occurrences of RR interval pairs, a histographic Poincaré plot can be constructed. Presented by van den Berg et al., a bidimensional histographic plot was developed to identify possible clustering of RR interval pairs [22]. Each point of the histographic Poincaré plot is associated with the number of occurrences of RR interval pairs given by the coordinates $(RR(n-1), RR(n))$. Representing the points by gray scale according to the number of the RR occurrences, different clusters can be appreciated as shown in Figure 5.6.

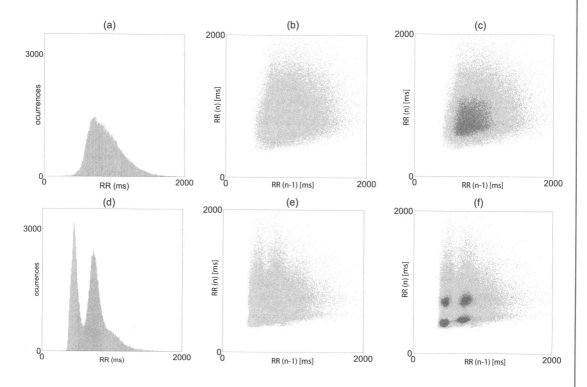

Figure 5.6: RR interval data from two patients with AF. (a),(d) 24-h RR interval histogram, (b),(e) Poincaré plot, and (c),(f) histographic Poincaré plot. Patient 1 presents unimodal RR histogram (top row), whereas patient 2 presents a bimodal one (bottom row). The histographic information of the Poincaré plot visualizes different number of clusters.

5.4.3   APPLICATIONS

Different considerations should be done about the interpretation of clusters. The Poincaré plot analysis does not provide direct evidence of dual pathways [27]. It is unclear whether the two pathway properties can be evaluated from the two populations of the Poincaré plot and the role of the second pathway in the conduction curve needs to be clarified [18, 32, 2]. However, the analysis of the Poincaré plot may be useful for evaluating drugs effects and AV node modifications on the ventricular response, even if the connection between different pathways and RR interval populations has not been completely understood yet.

By means of visual inspection of the histographic Poincaré plots, van der Berg et al. found RR clustering in 31 (47%) of the 66 patients under study [22]. Moreover, in patients with clustered RR intervals, electrical cardioversion resulted more often in restoration of SR. Also, those patients had a higher likelihood of remaining in SR than their counterparts without well-defined clusters. It was speculated that RR interval clustering represents a relatively high degree of AF organization.

5.5   TIME DOMAIN PARAMETERS

Time domain parameters can be computed following the recommendations for heart rate variability (HRV) measurements [33] that have also been applied during AF [34]. Time domain analysis of AF recordings usually includes the mean and standard deviation of normal-to-normal (SDNN) intervals, root mean-square differences of successive normal-to-normal intervals (rMSSD), and percentage of interval differences of successive normal-to-normal intervals greater than 50 ms (pNN50).

Time domain parameters are widely applied, e.g., to evaluate the effect of drugs [35, 36, 37], exercise [35, 38, 39, 40], and to predict outcome of cardioversion or invasive procedure results [41]. For example, SDNN showed a significant positive correlation with the exercise capacity and was the only independent predictor of normal exercise capacity [39] in patients with idiopathic AF. Effects of various drugs have been tested by time domain parameters, showing expected drugs characteristics [36, 37] as well as drug properties not commonly tested [35]. In particular, ECGs were recorded in a four-phase protocol (before and after infusion of flecainide, a commonly used antiarrhythmic drug, at rest and immediately after exercise) in order to characterize ventricular response to both exercise as autonomic nervous system (ANS) stimulus and flecainide which is also supposed to have effects on the ANS [35]. The results concerning ventricular response during exercise underlined the relevant activity played by the ANS in patients with AF, as time domain parameters decreased significantly. A pronounced decrease of rMMSD and pNN50, both highly related to vagal activity, highlighted the vagolitic effect of flecainide; see Figure 5.7.

5.6   SPECTRAL ANALYSIS AND $1/f$ POWER LAW BEHAVIOR

Atrial fibrillation is generally considered to be an exclusion criterion for spectral analysis of HRV because of the apparent total irregularity of ventricular rhythm during the arrhythmia. In particular, whereas time domain parameters can be easily computed in AF patients, frequency domain analysis has not been consistently performed as a consequence of the difficulty to interpret the huge number of spectral peaks of the variability signal. These considerations of complex determinants of ventricular rate during AF may explain why spectral analysis of RR intervals has mostly been applied to evaluate autonomic modulation of sinus node during SR before AF onset and after cardioversion rather than during AF episodes. Signs of enhanced parasympathetic or sympathetic activation have been reported by different authors [42, 43, 44, 45, 46, 47] who analyzed short-term HRV before paroxysmal

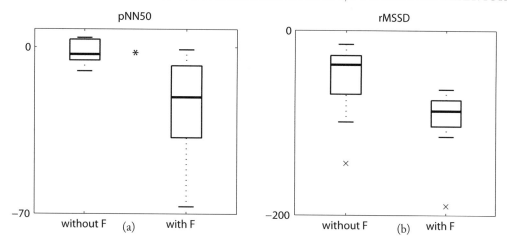

Figure 5.7: Box plots of the differences exercise-rest with and without flecainide (F) for (a) pNN50 and (b) rMSSD parameters. Each box plot has the following structure. The box itself contains the middle 50% of the data and the line in the box represents the median value of the data. The lower and upper edges of the box indicate the 25th and 75th percentile of the data set, respectively. Points at a greater distance from the median than 1.5 times the interquartile range are plotted individually as crosses. Note the pronounced exercise effect after flecainide. The asterisk indicates that the difference is statistically significant, $p < 0.05$.

AF onset. In particular, one common finding was the detection of transient alterations of neural modulatory activities before the arrhythmic event.

5.6.1   METHODOLOGICAL ASPECTS

Although classical spectral analysis is not commonly used in AF, the $1/f$ power law behavior has been successfully applied [48]. Briefly, log–log power spectrum has been reported to show linear decay with increasing frequency f, the spectral exponent β being defined as the value that satisfies the following equation:

$$S_x(f) \approx C\frac{1}{f^\beta}, \tag{5.6}$$

where $S_x(f)$ denotes power spectrum and C is a constant. By taking the logarithm of both sides of (5.6), one obtains

$$\log(S_x(f)) = \log C - \beta \log f, \tag{5.7}$$

showing that β can be estimated by linear regression analysis of $\log(S_x(f))$ on $\log f$. The power law relationship of RR interval variability is usually calculated in the frequency range of 10^{-4} to 10^{-2} Hz [49].

5.6.2   APPLICATIONS

The 24-h RR interval power spectrum during SR shows a $1/f$ noise-like downsloping linear pattern when plotted as log-power against log-frequency, while the same procedure applied in patients with chronic AF shows an angular shape with a breakpoint at which the spectrum is separated into two different regions [48]. Thus, the spectral characteristics revealed the presence of two correlation ranges: a white noise flat spectrum (no correlation) is observed at higher frequencies (>0.005 Hz), whereas the long-term spectral components (long-range correlation) are characterized by a $1/f$ power law behavior (Figure 5.8). The long-range correlation is similar to that observed in normal SR, suggesting that, during AF, long-term regulatory mechanisms are still effective in modulating the ventricular response. These findings are in agreement with observations on circadian rhythms in patients with chronic AF [50].

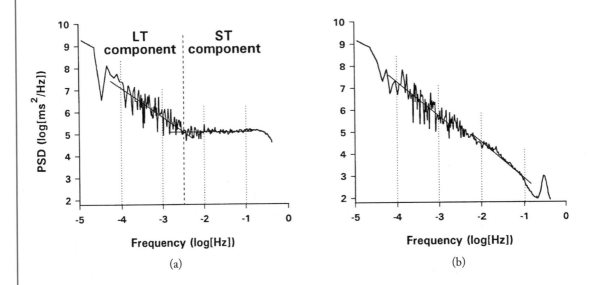

Figure 5.8:   Log–log power spectrum of 24-h RR fluctuations in (a) a representative patient with AF, and (b) a normal subject with SR. Vertical and horizontal axes have logarithmic scales. The vertical dashed lines in (a) represents estimated breakpoint frequency separating long-term (LT) and short-term (ST) components. (Reprinted from [48] with permission.)

5.7   NONLINEAR INDICES

5.7.1   EMBEDDING TIME SERIES DERIVED METHODS

The behavior of every nonrandom (deterministic) system can be considered to occur within a phase space. A topological equivalent of phase space can be constructed from empirical data (the RR interval series) using the technique of lags [51]. Choosing an embedding dimension L and a reconstruction delay τ ($\tau \geq 1$), the points that constitute a phase space are represented as the vector:

$$\mathbf{x}_\tau^L(n) = \begin{bmatrix} x(n) & x(n-\tau) & \cdots & x(n-(L-1)\tau) \end{bmatrix}, \tag{5.8}$$

where $x(n)$ denotes RR interval. For the sake of compactness, the subscript in (5.8) is omitted when $\tau = 1$. The technique is based on the assumption that the system generating the time series can be modeled as a low-order deterministic system.

Applications
Starting from the phase space, Stein et al. implemented an algorithm that uses nonlinear predictive forecasting for the RR interval series, predicting its future behavior for a few beats by observing other sufficiently similar trajectories in phase space [5]; see also [52]. Briefly, they reconstructed eight different phase spaces with embedding dimensions from 3–10. For each RR interval $x(i)$, the three nearest neighbors in each phase space are determined with the Cartesian distance metric,

$$d_{i,j} = \sqrt{\mathbf{x}^L(i) - \mathbf{x}^L(j))^2}, \tag{5.9}$$

chosen from every other point $x(j)$ in the same time series. The predicted next RR interval $x(i+1)$ is determined as the average of the RR intervals that follow the three nearest neighbors (each weighted by the inverse of its $d_{i,j}$). The Pearson correlation coefficient between predicted and actual RR intervals is computed for all intervals in the series for each of the eight phase spaces. The phase space with the embedding dimension yielding the highest correlation coefficient is thus determined. Within this embedding dimension, the three nearest neighbors of each RR interval trajectory in the phase space are used to predict the evolution of that trajectory from 1–10 steps into the future using a method analogous to that described in the second step. In their study, some RR interval series during AF presented predictability on very short-term scale. This weak predictability was suggested to represent the effect of cyclic oscillations in vagal and/or sympathetic tone at the level of the AV node.

5.7.2 ENTROPY

Approximate entropy (*ApEn*) is a measure of signal randomness, quantifying the unpredictability of fluctuations in a time series [53]. A set of L-length patterns is constructed from the RR interval series, taking L equally spaced RR intervals, cf. (5.8), and the correlation sum computed

$$C_\tau^L(r) = \frac{1}{N^2} \sum_{i,j=1; i \neq j}^{N} \Theta(r - d_{i,j,\tau}), \tag{5.10}$$

where $\Theta(\cdot)$ is the Heaviside step function and N is the number of possible L-length patterns. The distance $d_{i,j,\tau}$ is defined by

$$d_{i,j,\tau} = \max \left| \mathbf{x}_\tau^L(i) - \mathbf{x}_\tau^L(j) \right| \tag{5.11}$$

and is a similarity measure. The parameter r corresponds to the distance within which neighboring points (belonging to similar patterns) must lie and $C_\tau^L(r)$ expresses the prevalence of repetitive patterns of length L in the series. Finally, *ApEn* is defined as

$$ApEn(L, r, \tau) = \ln\left[\frac{C_\tau^L(r)}{C_\tau^{L+1}(r)}\right], \tag{5.12}$$

i.e., as the natural logarithm of the relative prevalence of repetitive patterns of length L compared with those of length $L + 1$. *ApEn* reflects the likelihood that similar patterns of observations will not be followed by additional similar observations. A time series containing many repetitive patterns, i.e., a regular and predictable series, has a relatively small *ApEn*; a less predictable, i.e., more complex, process has a larger *ApEn*.

Finally, the *ApEn* algorithm counts each sequence as matching itself, a practice carried over to avoid the occurrence of $\ln(0)$ in the calculations. This step has led to discussion of the bias of *ApEn*. Therefore, the sample entropy (*SampEn*), not counting self-matches, has been introduced [54].

Applications

A reduction in *ApEn* [54] or *SampEn* [55] has been considered as a hallmark of altered heart rate dynamics preceding spontaneous onset of AF. Also, *ApEn* has been found to have prognostic value in patients with chronic AF [56]. In particular, Yamada et al. computed *ApEn* of beat-to-beat and minute-to-minute fluctuations, denoted $ApEn_{b-b}$ and $ApEn_{m-m}$, respectively, of 24-h ambulatory ECGs in a population of 107 patients with chronic AF. Multivariate Cox models revealed that the best independent predictive value was obtained for $ApEn_{b-b}$. The parameter $ApEn_{b-b}$ could be modified by many factors. For example, atrial expansion and, thereby, induced reflex vagal excitation may increase the dispersion of atrial refractoriness and ventricular response irregularity. The atrial electrical remodeling induced by AF itself may shorten atrial refractoriness, thereby increase the number and frequency of f-waves, which could increase ventricular response irregularity by enhancing concealed conduction within the AV node. Although atrial expansion and electrical remodeling might progress as a pathological process of chronic AF, it was observed that neither left atrial diameter nor the duration of chronic AF had prognostic value; furthermore, these factors, if involved, would increase ventricular response irregularity.

5.7.3 REGULARITY

An index of regularity, measuring the degree of recurrence of a pattern in a signal, can be defined by means of the conditional entropy (*CE*). The index *CE* represents the amount of information carried by the most recent sample of a series $\mathbf{x}$ when its past $L - 1$ samples are known [57]. For a give process $\mathbf{x} = \{x(n), n = 1, 2, \ldots, N\}$ and a L-length sequence $\mathbf{x}^L(n)$ extracted from $\mathbf{x}$, CE is defined as:

$$CE(L) = E(\mathbf{x}^L) - E(\mathbf{x}^{L-1}) = -\sum p(\mathbf{x}^L)\log \mathbf{x}^L + \sum p(\mathbf{x}^{L-1})\log \mathbf{x}^{L-1}, \tag{5.13}$$

where $E(\cdot)$ is the Shannon entropy and $p(\cdot)$ is the probability of encountering a certain pattern. Summation is extended to all the possible L-length patterns of $\mathbf{x}$. Equation (5.13) emphasizes that CE measures the amount of information obtained when the pattern length is augmented from $L-1$ to L. If a process is periodic (i.e., perfectly predictable) and has been observed for a sufficient time, we will be able to predict the next samples. Therefore, there will be no increase of information by increasing the pattern length and CE will go to zero after a certain L. Equation (5.13) needs the computation of probabilities p, which are generally unknown, and are estimated using sample probability. Unfortunately, the pattern length should be large enough to produce a robust estimate of sample probability, which becomes unreliable when L reaches the number of available sample N. In the latter case, the existence of several patterns observed only once drops CE to zero even for a random process. To compensate for a limited number of sample, the corrected conditional entropy (CCE) must be used

$$CCE(L) = CE(L) + \text{perc}(L) \cdot E(\mathbf{x}^1),\qquad(5.14)$$

where $\text{perc}(L)$ is the percentage of single points (i.e., patterns encountered only once) in the L-dimensional space. As this term increases with L, it tends to compensate the decreases of $CE(L)$ induced by the pattern found only once in $\mathbf{x}$. Therefore, $CCE(L)$ is the sum of two contributions, one decreasing and the other increasing with L. A regularity index is therefore defined as

$$R = 1 - \min\left(\frac{CCE}{E(\mathbf{x}^1)}\right),\qquad(5.15)$$

where CCE is normalized by the Shannon entropy of the process to derive an index which is independent of the different probability distribution of the processes. The index R tends to 1 when the process $\mathbf{x}$ is periodic, i.e., perfectly predictable, falls to zero in the presence of high complexity signals, i.e., completely unpredictable, and remains in between for intermediate processes.

Applications

The regularity index R has recently been applied to the evaluation of exercise effect on ECG recordings from patients with persistent AF [35]. As the ANS plays an important role among the factors influencing ventricular response by modulating refractoriness of the AV node, that is mainly dependent on vagal tone [58], the purpose of the study was to characterize ventricular response during AF to changes of the autonomic balance induced by exercise. The R index, reflecting nonlinear series predictability, tended to increase during exercise. It was found that R values are very low compared to SR [59], thus the predictability degree of ventricular response is very small during AF. Nevertheless, taking linear and nonlinear dynamics into account, R succeeded in underlining the increased predictability of ventricular response during exercise. The results highlighted the relevant activity played by the ANS in patients with AF, as time domain parameters decreased and predictability indices increased. In addition, R has been used in order to assess different characteristics of termination, i.e., AF that terminates immediately (within 1 s), and nonterminating AF episodes, i.e., AF that was not observed to terminate for the duration of the long-term recording, at least for an hour.

In particular, R succeeded in discriminating between nonterminating AF and terminating episodes ($R_{N-T} = 0.181 \pm 0.098$ versus $R_T = 0.104 \pm 0.056$, $p < 0.05$) [60].

5.8   CONCLUSIONS

Many factors are important in determining the ventricular response and irregularity as a result of rapid and variable atrial rate. The findings of the various studies, briefly illustrated in this chapter, strengthen the opinion that the RR interval behavior does not represent a random AV condition but rather reflects the electrophysiological properties of the AV node. However, different theories about pathophysiological meaning and clinical application are still under investigation. Signal processing plays an essential role for a deeper understanding of AF mechanisms and for quantifying AF patterns, thus providing valuable information for the selection and evaluation of a specific treatment. Future studies should shed light on this complex electrical conduction system.

Bibliography

[1] A. C. Rankin and A. J. Workman, "Rate control in atrial fibrillation: role of atrial inputs to the AV node," *Cardiovasc. Res.*, vol. 44, pp. 249–251, 1999. DOI: 10.1016/S0008-6363(99)00248-5

[2] Y. H. Zhang and T. N. Mazgalev, "Ventricular rate control during atrial fibrillation and av node modifications: past, present, and future," *Pacing Clin. Electrophysiol.*, vol. 27, pp. 382–393, 2004. DOI: 10.1111/j.1540-8159.2004.00447.x

[3] F. J. Chorro, C. J. Kirchhof, J. Brugada, and M. A. Allessie, "Ventricular response during irregular atrial pacing and atrial fibrillation," *Am. J. Physiol.*, vol. 259, pp. H1015–H1021, 1990.

[4] A. Bollmann and F. Lombardi, "Electrocardiology of atrial fibrillation—Current knowledge and future challenges," *IEEE Eng. Med. Biol. Mag.*, vol. 25, pp. 15–23, 2006. DOI: 10.1109/EMB-M.2006.250504

[5] K. M. Stein, J. Waldeen, N. Lippman, and B. B. Lerman, "Ventricular response in atrial fibrillation: random or deterministic?," *Am. J. Physiol. (Heart Circ. Physiol.)*, vol. 277, pp. H452–H458, 1999.

[6] S. B. Olsson, N. Cai, M. Dohnal, and K. K. Talwar, "Noninvasive support for and characterization of multiple intranodal pathways in patients with mitral valve disease and atrial fibrillation," *Eur. Heart J.*, vol. 7, pp. 320–333, 1986.

[7] S. Rokas, S. Gaitanidou, S. Chatzidou, N. Agrios, and S. Stamatelopoulos, "A noninvasive method for the detection of dual atrioventricular node physiology in chronic atrial fibrillation," *Am. J. Cardiol.*, vol. 84, pp. 1442–1445, 1999. DOI: 10.1016/S0002-9149(99)00593-7

[8] S. Rokas, S. Gaitanidou, S. Chatzidou, C. Pamboucas, D. Achtipis, and S. Stamatelopoulos, "Atrioventricular node modification in patients with chronic atrial fibrillation—Role of morphology of RR interval variation," *Circulation*, vol. 103, pp. 2942–2948, 2001. DOI: 10.1016/S1062-1458(01)00499-8

[9] J. Tebbenjohanns, B. Schumacher, T. Korte, M. Niehaus, and D. Pfeiffer, "Bimodal RR interval distribution in chronic atrial fibrillation: Impact of dual atrioventricular nodal physiology on long–term rate control after catheter ablation of the posterior atrionodal input," *J. Cardiovasc. Electrophysiol.*, vol. 11, pp. 497–503, 2000. DOI: 10.1111/j.1540-8167.2000.tb00001.x

[10] S. Gaitanidou, S. Rokas, C. Pamboucas, D. Actipis, S. Chatzidou, J. Darsinos, S. Stamatelopou-los, and S. Moulopoulos, "The RR interval distribution pattern as a predictive factor of the outcome of atrioventricular conduction modification in patients with atrial fibrillation," *J. Am. Coll. Cardiol.*, vol. 31, pp. 333A–334A, 1998. DOI: 10.1016/S0735-1097(97)85166-2

[11] F. Hegbom, O. M. Orning, M. Heldal, and K. Gjesdal, "Effects of ablation, digi-talis, and beta–blocker on dual atrioventricular nodal pathways and conduction dur-ing atrial fibrillation," *J. Cardiovasc. Electrophysiol.*, vol. 15, pp. 1141–1146, 2004. DOI: 10.1046/j.1540-8167.2004.04014.x

[12] N. Cai, M. Dohnal, and S. B. Olsson, "Methodological aspects of the use of heart rate stratified RR interval histograms in the analysis of atrioventricular conduction during atrial fibrillation," *Cardiovasc. Res.*, vol. 21, pp. 455–462, 1987. DOI: 10.1093/cvr/21.6.455

[13] J. Carlson, *Exploration of supraventricular conduction with respect to atrial fibrillation*. PhD thesis, Faculty of Medicine, Lund University, Sweden, 2005.

[14] M. P. Ingemansson, J. Carlson, and S. B. Olsson, "Modification of intrinsic AV–nodal properties by magnesium in combination with glucose, insulin, and potassium (GIK) during chronic atrial fibrillation," *J. Electrocardiol.*, vol. 31, pp. 281–292, 1998. DOI: 10.1016/S0022-0736(98)90013-9

[15] J. Billette, R. A. Nadeau, and F. Roberge, "Relation between minimum RR interval during atrial fibrillation and functional refractory period of AV junction," *Cardiovasc. Res.*, vol. 8, pp. 347–351, 1974. DOI: 10.1093/cvr/8.3.347

[16] A. U. Khand, A. C. Rankin, J. G. F. Cleland, I. Gemmell, E. Clark, and P. W. Macfarlane, "The assessment of autonomic function in chronic atrial fibrillation: Description of a noninvasive technique based on circadian rhythm of atrioventricular nodal functional refractory periods," *Europace*, vol. 8, pp. 927–934, 2006. DOI: 10.1093/europace/eul111

[17] R. J. Cohen and R. D. Berger, "A quantitative model for the ventricular response during atrial fibrillation," *IEEE Trans. Biomed. Eng.*, vol. 30, pp. 769–781, 1983. DOI: 10.1109/TBME.1983.325077

[18] Y. H. Zhang, K. A. Mowrey, and T. Mazgalev, "Does RR interval distribution during atrial fib-rillation indicate dual pathway atrioventricular nodal electrophysiology?," *Circulation*, vol. 106, p. 179, 2002.

[19] T. Anan, K. Sunagawa, H. Araki, and M. Nakamura, "Arrhythmia analysis by successive RR plotting," *J. Electrocardiol.*, vol. 23, pp. 243–248, 1990. DOI: 10.1016/0022-0736(90)90163-V

[20] A. S. Chishaki, K. Sunagawa, K. Hayashida, M. Sugimachi, and M. Nakamura, "Identification of the rate–dependent functional refractory period of the atrioventricular node in simulated atrial fibrillation," *Am. Heart J.*, vol. 121, pp. 820–826, 1991. DOI: 10.1016/0002-8703(91)90194-M

[21] A. C. Suyama, K. Sunagawa, M. Sugimachi, T. Anan, K. Egashira, and A. Takeshita, "Differentiation between aberrant ventricular conduction and ventricular ectopy in atrial fibrillation using RR interval scattergram," *Circulation*, vol. 88, pp. 2307–2314, 1993.

[22] M. P. van den Berg, T. van Noord, J. Brouwer, J. Haaksma, D. J. van Veldhuisen, H. J. G. M. Crijns, and I. C. van Gelder, "Clustering of RR intervals predicts effective electrical cardioversion for atrial fibrillation," *J. Cardiovasc. Electrophysiol.*, vol. 15, pp. 1027–1033, 2004. DOI: 10.1046/j.1540-8167.2004.03686.x

[23] A. Bollmann, D. Husser, L. Mainardi, F. Lombardi, P. Langley, A. Murray, J. J. Rieta, J. Millet, S. B. Olsson, M. Stridh, and L. Sörnmo, "Analysis of surface electrocardiograms in atrial fibrillation: Techniques, research, and clinical applications," *Europace*, vol. 8, pp. 911–926, 2006. DOI: 10.1093/europace/eul113

[24] A. M. Climent, D. Husser, V. D. A. Corino, L. Mainardi, H. U. Klein, J. Millet, and A. Bollmann, "Non-invasive assessment of atrioventricular conduction properties and their effects on ventricular response in atrial fibrillation," in *Proc. Comput. Cardiol.*, vol. 33, pp. 105–108, http://cinc.mit.edu, 2006.

[25] J. Hayano, S. Sakata, A. Okada, S. Mukai, and T. Fujinami, "Circadian rhythms of atrioventricular conduction properties in chronic atrial fibrillation with and without heart failure," *J. Am. Coll. Cardiol.*, vol. 31, pp. 158–166, 1998. DOI: 10.1016/S0735-1097(97)00429-4

[26] J. Hayano, S. Ishihara, H. Fukuta, S. Sakata, S. Mukai, N. Ohte, and G. Kimura, "Circadian rhythm of atrioventricular conduction predicts long–term survival in patients with chronic atrial fibrillation," *Chronobiol. Int.*, vol. 19, pp. 633–648, 2002. DOI: 10.1081/CBI-120004223

[27] T. Oka, T. Nakatsu, S. Kusachi, Y. Tominaga, S. Toyonaga, H. Ohnishi, M. Nakahama, I. Komatsubara, M. Murakami, and T. Tsuji, "Double–sector Lorenz plot scattering in an R–R interval analysis of patients with chronic atrial fibrillation – incidence and characteristics of vertices of the double–sector scatteringsector Lorenz plot scattering in an R–R interval analysis of patients with chronic atrial fibrillation—Incidence and characteristics of vertices of the double–sector scattering," *J. Electrocardiol.*, vol. 31, pp. 227–235, 1998. DOI: 10.1016/S0022-0736(98)90138-8

[28] A. C. Suyama, "Aberrant ventricular conduction and ventricular ectopy—Reply," *Circulation*, vol. 89, pp. 2945–2945, 1994.

[29] A. M. Climent, M. S. Guillem, D. Husser, F. J. Castells, J. Millet, and A. Bollmann, "Circadian rhythm of ventricular response during atrial fibrillation is not determined by dual atrioventricular nodal pathway conduction," *J. Electrocardiol.*, vol. 40, p. S30, 2007. DOI: 10.1016/j.jelectrocard.2007.08.005

[30] P. V. C. Hough, "Method and means for recognizing complex patterns," *US Patent*, vol. 3,069,654, 1962.

[31] R. Duda and P. Hart, "Use of the Hough transformation to detect lines and curves in pictures," *Commun. ACM*, vol. 15, pp. 11–15, 1972. DOI: 10.1145/361237.361242

[32] Y. H. Zhang, S. Bharati, K. A. Mowrey, and T. N. Mazgalev, "His electrogram alternans reveal dual atrioventricular nodal pathway conduction during atrial fibrillation— The role of slow pathway modification," *Circulation*, vol. 107, pp. 1059–1065, 2003. DOI: 10.1161/01.CIR.0000051464.52601.F4

[33] A. J. Camm, M. Malik, J. T. Bigger, G. Breithardt, S. Cerutti, R. J. Cohen, P. Coumel, E. L. Fallen, H. L. Kennedy, R. E. Kleiger, F. Lombardi, A. Malliani, A. J. Moss, J. N. Rottman, G. Schmidt, P. J. Schwartz, and D. Singer, "Heart rate variability—Standards of measurement, physiological interpretation, and clinical use," *Circulation*, vol. 93, pp. 1043–1065, 1996.

[34] M. P. van den Berg, J. Haaksma, J. Brouwer, R. G. Tieleman, G. Mulder, and J. G. M. Crijns, "Heart rate variability in patients with atrial fibrillation is related to vagal tone," *Circulation*, vol. 96, pp. 1209–1216, 1997.

[35] V. D. A. Corino, L. T. Mainardi, D. Husser, H. U. Klein, and A. Bollmann, "Ventricular response during atrial fibrillation: Evaluation of exercise and flecainide effects," in *Proc. Comput. Cardiol.*, vol. 33, pp. 145–148, http://cinc.mit.edu, 2006.

[36] M. H. Hsieh, S. A. Chen, Z. C. Wen, C. T. Tai, C. E. Chiang, Y. A. Ding, and M. S. Chang, "Effects of antiarrhythmic drugs on variability of ventricular rate and exercise performance in chronic atrial fibrillation complicated with ventricular arrhythmias," *Int. J. Cardiol.*, vol. 64, pp. 37–45, 1998. DOI: 10.1016/S0167-5273(97)00330-6

[37] A. Incze, A. Frigy, and S. Cotoi, "The efficacy of sublingual verapamil in controlling rapid ventricular rate in chronic atrial fibrillation," *Rom. J. Intern. Med.*, vol. 36, pp. 219–225, 1998.

[38] O. Akyurek, E. Diker, M. Guldal, and D. Oral, "Predictive value of heart rate variability for the recurrence of chronic atrial fibrillation after electrical cardioversion," *Clin. Cardiol.*, vol. 26, pp. 196–200, 2003. DOI: 10.1002/clc.4960260411

[39] M. Matsumoto, T. Yamashita, E. Fukuda, K. Sagara, H. Linuma, and L. T. Fu, "Relation between variability of ventricular response intervals and exercise capacity in patients with non–valvular atrial fibrillation," *Circ. J.*, vol. 68, pp. 294–296, 2004. DOI: 10.1253/circj.68.294

[40] O. Piot, C. Chauvel, A. Lazarus, D. Pellerin, D. David, L. Leneveut-Ledoux, L. Guize, and J. Y. Le Heuzey, "Effects of a selective A1-adenosine receptor agonist on heart rate and heart rate variability during permanent atrial fibrillation," *Pacing Clin. Electrophysiol.*, vol. 21, pp. 2459–2464, 1998. DOI: 10.1111/j.1540-8159.1998.tb01201.x

[41] S. Lönnerholm, P. Blomström, L. Nilsson, M. Ericson, M. Kesek, L. Jidéus, and C. Blomström-Lundqvist, "Autonomic denervation after the maze procedure," *Pacing Clin. Electrophysiol.*, vol. 26, pp. 587–592, 2003. DOI: 10.1046/j.1460-9592.2003.00098.x

[42] M. Bettoni and M. Zimmermann, "Autonomic tone variations before the onset of paroxysmal atrial fibrillation," *Circulation*, vol. 105, pp. 2753–2759, 2002. DOI: 10.1161/01.CIR.0000018443.44005.D8

[43] C. Dimmer, R. Tavernier, N. Gjorgov, G. Van Nooten, D. L. Clement, and L. Jordaens, "Variations of autonomic tone preceding onset of atrial fibrillation after coronary artery bypass grafting," *Am. J. Cardiol.*, vol. 82, pp. 22–25, 1998. DOI: 10.1016/S0002-9149(98)00231-8

[44] B. Herweg, P. Dalal, B. Nagy, and P. Schweitzer, "Power spectral analysis of heart period variability of preceding sinus rhythm before initiation of paroxysmal atrial fibrillation," *Am. J. Cardiol.*, vol. 82, pp. 869–874, 1998.

[45] F. Lombardi, D. Tarricone, F. Tundo, F. Colombo, S. Belletti, and C. Fiorentini, "Autonomic nervous system and paroxysmal atrial fibrillation: a study based on the analysis of RR interval changes before, during and after paroxysmal atrial fibrillation," *Eur. Heart J.*, vol. 25, pp. 1242–1248, 2004. DOI: 10.1016/j.ehj.2004.05.016

[46] Z. C. Wen, S. A. Chen, C. T. Tai, J. L. Huang, and M. S. Chang, "Role of autonomic tone in facilitating spontaneous onset of typical atrial flutter," *J. Am. Coll. Cardiol.*, vol. 31, pp. 602–607, 1998. DOI: 10.1016/S0735-1097(97)00555-X

[47] Y. Tomoda, S. Uemura, S. Fujimoto, H. Yamamoto, T. Hashimoto, and K. Dohi, "Assessment of autonomic nervous activity before the onset of paroxysmal atrial fibrillation," *J. Cardiol.*, vol. 31, pp. 11–17, 1998.

[48] J. Hayano, Y. Sakakibara, M. Yamada, N. Ohte, T. Fujinami, K. Yokoyama, Y. Watanabe, and K. Takata, "Decreased magnitude of heart-rate spectral components in coronary artery disease—Its relation to angiographic severity," *Circulation*, vol. 81, pp. 1217–1224, 1990.

[49] J. T. Bigger Jr., R. C. Steinman, L. M. Rolnitzky, J. L. Fleiss, P. Albrecht, and R. J. Cohen, "Power law behavior of RR-interval variability in healthy middle-aged persons, patients with recent acute myocardial infarction, and patients with heart transplants," *Circulation*, vol. 93, pp. 2142–2151, 1996.

[50] E. A. Raeder, "Circadian fluctuations in ventricular response to atrial fibrillation," *J. Am. Coll. Cardiol.*, vol. 15, pp. 1200–1200, 1990.

[51] F. Takens, "Detecting strange attractors in turbulence," *Lect. Notes Math.*, vol. 898, pp. 366–381, 1980. DOI: 10.1007/BFb0091924

[52] G. Sugihara and R. M. May, "Nonlinear forecasting as a way of distinguishing chaos from measurement error in time–series," *Nature*, vol. 344, pp. 734–741, 1990. DOI: 10.1038/344734a0

[53] S. M. Pincus, "Approximate entropy as a measure of system–complexity," *Proc. Natl. Acad. Sci. USA*, vol. 88, pp. 2297–2301, 1991. DOI: 10.1073/pnas.88.6.2297

[54] J. S. Richman and J. R. Moorman, "Physiological time series analysis using approximate entropy and sample entropy," *Am. J. Physiol. (Heart Circ. Physiol.)*, vol. 278, pp. H2039–H2049, 2000.

[55] V. Tuzcu, S. Nas, T. Borklu, and A. Ugur, "Decrease in the heart rate complexity prior to the onset of atrial fibrillation," *Europace*, vol. 8, pp. 398–402, 2006. DOI: 10.1093/europace/eul031

[56] A. Yamada, J. Hayano, S. Sakata, A. Okada, S. Mukai, N. Ohte, and G. Kimura, "Reduced ventricular response irregularity is associated with increased mortality in patients with chronic atrial fibrillation," *Circulation*, vol. 102, pp. 300–306, 2000.

[57] A. Porta, G. Baselli, D. Liberati, N. Montano, C. Cogliati, T. Gnecchi-Ruscone, A. Malliani, and S. Cerutti, "Measuring regularity by means of a corrected conditional entropy in sympathetic outflow," *Biol. Cybern.*, vol. 78, pp. 71–78, 1998. DOI: 10.1007/s004220050414

[58] L. Toivonen, A. Kadish, W. Kou, and F. Morady, "Determinants of the ventricular rate during atrial fibrillation," *J. Am. Coll. Cardiol.*, vol. 16, pp. 1194–1200, 1990.

[59] L. T. Mainardi, M. Matteucci, and R. Sassi, "On predicting the spontaneous termination of atrial fibrillation episodes using linear and nonlinear parameters of ECG signal and RR series.," in *Proc. Comput. Cardiol.*, vol. 31, pp. 665–668, IEEE Press, 2004. DOI: 10.1109/CIC.2004.1443026

[60] L. T. Mainardi, V. D. A. Corino, L. Lombardi, C. Tondo, M. Mantica, F. Lombardi, and S. Cerutti, "Assessment of the dynamics of atrial signals and local atrial period series during atrial fibrillation: effects of isoproterenol administration.," *Biomed. Eng. Online*, vol. 3, p. 37, 2004. DOI: 10.1186/1475-925X-3-37

CHAPTER 6

Organization Measures of Atrial Activity During Fibrillation

Flavia Ravelli, Luca Faes, Valentina Corino, and Luca Mainardi

6.1 INTRODUCTION

A growing body of experimental and clinical observations reveals, under the apparently random activation pattern of atrial fibrillation (AF), the presence of an underlying order sustained by numerous factors such as myocardium anatomy, electrophysiological properties, and autonomic innervation [1, 2, 3]. To grow and enforce this assertion, a variety of signal processing methods has been applied to analyze atrial recordings and to characterize the *level of organization* of the fibrillating atrium [4, 5]. The derived measures are becoming of great importance for electrophysiologists, because they convey relevant physiological and clinical information. For example, indices of organization may be related to the electrophysiological mechanisms sustaining AF, or may help to evaluate new strategies for the treatment of AF such as catheter ablation or electrical cardioversion. Moreover, the spatiotemporal mapping of organization may aid to recognize the patterns of wave propagation and to evaluate their correlation with atrial functional and structural characteristics.

The present chapter summarizes the state-of-art of how to quantify the degree of atrial organization during AF through a review of the main signal processing techniques employed for the analysis of atrial electrical activity. As the concept of atrial organization may assume different meanings in the context of AF, particular attention is paid to stress those peculiar characters of organization probed and captured by each method.

6.2 THE CONCEPT OF AF ORGANIZATION

A unified definition of *AF organization* is not available and the term has been used to describe various characteristics of electrical activity of the fibrillating atria, such as the repeatability/regularity of the atrial activations, the correlation/synchronicity among electrograms recorded in different locations, or the similarity of the wave morphology. From a signal processing viewpoint, the lack of a unique definition generated a variety of measures in attempts to capture signal features derived from subjective interpretation of this nebulous concept. As a consequence, there is an obvious difficulty both to compare results from different studies (and algorithms), and to give interpretations to novel indices.

In the light of these considerations, the need to provide a categorization of the various measurements has been debated [4, 5]. Indeed, many possible criteria to categorize the methods exist, including the number of recording sites, the signal processing domains (time versus frequency domain or linear versus nonlinear domain), or the finality of the methods (understanding AF, planning treatment or therapy, etc.). As many of the papers facing the problem of quantifying AF organization attempt to interpret the signal processing indices in terms of electrophysiological properties or reflected AF mechanisms, we believe that a useful categorization method should be based on the electrophysiological features which are investigated as a hallmark of organization. Following this line, we categorized algorithms in four classes in this chapter. Each class includes methods which share the conceptual interpretation of organization in terms of:

1. temporal regularity (rhythmic) of atrial activations,

2. regularity/complexity of single site electrograms,

3. coupling/synchronization between the electrical activity of two adjacent sites, and

4. similarity of activation wave morphologies.

6.3   MEASURES OF AF ORGANIZATION

6.3.1   RHYTHM ANALYSIS

The most intuitive way to characterize the atrial rhythm is to analyze the sequence of the atrial periods, i.e., the time intervals between the occurrences of two consecutive local activation waves. Simple statistics, such as the mean value and the coefficient of variation of the atrial period sequence, may be related to the specific electrophysiological mechanism underlying the arrhythmia. During multiple wavelet propagation, it was shown that more disorganized AF episodes are associated with atrial cycles of shorter duration and higher beat-to-beat variability [3, 6]. During AF triggered by pulmonary vein ectopic activity, sites of shorter cycle length have been found as good ablation targets [7], while atrial period lengthening is a good predictor of ablation success [8].

Besides these simple calculations, the sequence of atrial periods can be characterized through more sophisticated time series analysis in order to investigate the presence of complex dynamics. With this approach, Mainardi et al. [9] introduced both linear and nonlinear parameters to study the dynamics underlying the atrial period variability. The adopted indices were the level of predictability (LP) and the regularity index R; the latter index was previously defined in Chapter 5 in the context of ventricular response characterization. The index LP is linear, defined as the percentage of signal power which can be explained by prediction when the series is modeled by an autoregressive model of order p,

$$x(n) = \sum_{k=1}^{p} a_k x(n-k) + \eta(n) , \tag{6.1}$$

where the a_k's are the model coefficients, and $\eta(n)$ is a white noise process with variance σ_η^2 feeding the model. The actual sample $x(n)$ differs from its model prediction $\hat{x}(n)$, thus generating the prediction error

$$e(n) = x(n) - \hat{x}(n) = x(n) - \sum_{k=1}^{p} a_k x(n-k) . \tag{6.2}$$

The index LP is then defined as follows:

$$LP = 1 - \frac{\sigma_e^2}{\sigma_x^2} . \tag{6.3}$$

In (6.3), σ_x^2 represents the variance of the signal $x(n)$ and σ_e^2 the variance of the prediction error $e(n)$. These indices were used to perform an automatic discrimination of different atrial rhythms and AF types [9], and to capture the fine changing dynamics of the atrial period during modifications of the autonomic nervous system control induced by drug administration [10].

From a methodological viewpoint, the reliability of organization indices based on atrial cycle length depends on the precision by which the times of the local electrical activations of the atrial tissue are estimated. Whereas in unipolar recordings the time of the maximum signal slope is elective as it is closely related to local electrophysiological parameters [11], different criteria have been proposed for bipolar recordings, which are the most adopted in clinical practice [12]. Among them, the instance of the maximum of the signal and of its first derivative are commonly used to locate the activation time during AF episodes of low complexity [13, 14], while methods considering the wave morphology seem more reliable in the presence of complex and fragmented activations [15, 16].

As an alternative to time domain analysis of the atrial period, frequency domain analysis of atrial signals constitutes a valid approach to investigate the rhythm of AF electrograms. In this context, Skanes et al. [17] introduced the dominant frequency (DF) concept to identify spatially similar sources of periodic activity during complex patterns of activation seen in AF. Dominant frequency is defined as the peak frequency of the spectrum of AF bipolar electrograms within the 0.4–60 Hz band. Subsequent to this work, the DF has been successfully exploited in many electrophysiological studies, and was indicated as an important parameter for guiding clinical treatment of AF. The existence of a left-to-right gradient in the DF observed in animal models [18] supported the hypothesis that the left atrium plays an important role in sustaining AF. Moreover, the presence of the DF gradient in human paroxysmal AF [19], and its loss after isolation of the pulmonary veins [20], indicated the presence of elective regions for the ablative treatment. More generally, the study of the DF regional distribution in different types of AF [8] was helpful to identify critical areas for the maintenance of the arrhythmia, and consequently the target of ablation procedures.

Another frequency domain parameter reflecting the organization of AF was proposed by Everett et al. [21], who introduced an organization index derived from postprocessing of the AF electrogram spectrum. Briefly, the area under the DF peak and three of its harmonic is computed and the organization index is calculated as the ratio between this value and the total area of the spectrum evaluated from 2.5 Hz up to, but not including, the 5th harmonic peak. Even if no physiological explanation of the meaning of the index was provided, the authors showed in animal models that the organization index is related to the efficacy of attempts to terminate AF through defibrillation shocks [22] or burst pace delivery [23]. The index was also exploited in association with the information given by the DF to provide a full frequency domain characterization of AF electrograms, and to evaluate the corresponding informative content in relation to catheter ablation [24].

6.3.2   REGULARITY ANALYSIS

As an alternative to the investigation of specific rhythms, AF organization can be quantified from single atrial recordings by analysis of the whole signal aimed to infer measures related to the dynamical complexity of the signal itself. The rationale behind this approach is that features, such as the presence of undisturbed portions of the signal or the repetitiveness over time of similar patterns, are indicative of high regularity, or low dynamical complexity, related to the temporal organization of the arrhythmia. Within this viewpoint, many authors have proposed to quantify organization as the presence of a clear baseline, low correlation dimension, entropy or complexity, high predictability, or regularity in the analyzed atrial signals.

Barbaro et al. [25] proposed to use the number of occurrences, defined as the percentage of points along the baseline of a single atrial signal, as a linear time domain measure of the local organization during AF. This parameter was selected from a set of single-site measures investigated by the same authors [26] as the one that best classified AF signals according to the Wells' criteria [27], and was shown to have a good time resolution, to be robust to far-field artifacts, and to be easy to implement in real-time applications. Indeed, the high time resolution of the algorithm, associated with the good spatial resolution allowed by multipolar basket catheters, led the authors to evaluate the spatiotemporal distribution of organization in human AF, eliciting individual and well-defined regional patterns of organization that were supposed to have implications in the choice of the regions candidate for ablation [28].

Along with linear analysis, methods taken from the theory of nonlinear dynamic systems were implemented to describe the local electrical activity of the atrial tissue during fibrillation. After the intriguing studies addressing the issue of determining whether cardiac fibrillation contains deterministic features or not [29, 30], several authors have applied tools, provided from chaos theory and nonlinear dynamics, to AF signals. The study of Hoekstra et al. [31] was the first exhaustive nonlinear analysis of AF in man. The authors estimated the coarse-grained correlation dimension and the correlation entropy of unipolar epicardial electrograms acquired during various atrial rhythms, particularly during AF. The calculation of these two parameters is based on the time delay embedding procedure applied on the investigated signal by which delay vectors are constructed from the signal $x(n)$, taking m samples spaced in time by a delay τ, i.e., $\mathbf{x}(n) = [x(n), x(n - \tau), \ldots, x(n - (m - 1)\tau)]$. The correlation sum $C_m(r)$ is then estimated by counting the number of vectors $(\mathbf{x}(i), \mathbf{x}(j))$, whose distance is less than r in the m-dimensional space

$$C_m(r) = \frac{1}{N^2} \sum_{i,j=1}^{N} \Theta(r - \|\mathbf{x}(i) - \mathbf{x}(j)\|) , \qquad (6.4)$$

where Θ is the Heaviside function, N is the number of possible delay vectors, and $\|\cdot\|$ indicates a generic vector norm. In [31], the supremum norm, defined by

$$\|\mathbf{x}(i) - \mathbf{x}(j)\| = \max_{0 \leq k < m} |x(i + k) - x(j + k)| , \qquad (6.5)$$

was used for computational efficiency. In the same work, coarse–grained correlation dimension (CD) was computed as the local slope of the correlation sum,

$$CD = \frac{\partial \ln C_m(r)}{\partial \ln r} , \qquad (6.6)$$

and coarse–grained correlation entropy (CE) as

$$CE = \frac{1}{n\tau} \ln \frac{C_m(r)}{C_{m+n}(r)} \qquad (6.7)$$

with $n = 2$. The embedding dimension m and the distance r were optimized to the specific data to obtain coarse–grained estimates of CD and CE. These two parameters are related to the complexity of the dynamics in terms of dimensionality and divergence of trajectories of the chaotic attractor. In the study of Hoekstra et al. [31], they were exploited to discriminate among electrograms during electrically induced AF in humans, according to the classification scheme proposed by Konings et al. [3] for unipolar electrograms collected from high resolution epicardial mapping. The study revealed the presence of nonlinear dynamics in type I AF, while type II and type III AF did not appear to exhibit features of low-dimensional chaos.

Nonlinear analysis methods have also been applied to bipolar signals collected by basket catheter. In 1998, Pitschner et al. [32] calculated the correlation dimension of the depolarization wavefronts on signals measured during paroxysmal AF, and found that the area anterior to the tricuspid valve showed the most pronounced chaotic activity. In 1999, Berkowitsch et al. [33] proposed a combination of symbolic dynamics and adaptive power estimation to compute the normalized algorithmic complexity of single bipolar endocardial electrograms. This technique produces a measure of the "redundancies" in patterns of the AF electrogram so that the algorithmic complexity is inversely related to the number of redundancies found in the analyzed signal. The method was then used to show heterogeneous complexity among different atrial regions, and complexity changes after drug administration [34].

In the context of the analysis of single-site atrial recordings, a comparison between linear and nonlinear measures of the signals and of the local atrial period series was performed by Mainardi et al. [9, 10, 35]. The authors found that both linear and nonlinear indices (i.e., LP and R) were able to discriminate among different atrial rhythms, and particularly within different AF complexity classes [9, 35], and to capture subtle changes due to isoproterenol infusion both during normal sinus rhythm and AF [10].

6.3.3   SYNCHRONIZATION ANALYSIS

The concept of organization may be investigated from mutual analysis of pairs of recordings simultaneously collected during different atrial rhythms. In this case, measuring organization implies judging the electrical activity at one site in relation to the activity of another. Measures derived in such a way emphasize the concepts of relative temporal behavior and spatial coordination between electrical activations occurring at different sites. In respect to approaches developed for single-site recordings, the introduction of algorithms involving two (or more) signals may provide complementary information. For instance, synchronization measures could be exploited to investigate the preferential directions of waveform propagation during arrhythmias, or to reflect the spatial dispersion of electrophysiological parameters such as conduction velocity and refractory period. Some of these aspects have relevance when the distances between the recording sites are known and numerous electrodes are available.

An attempt to evaluate the degree of coupling between atrial electrograms was performed by Ropella et al. [36], who used a frequency domain method to discriminate between fibrillatory and nonfibrillatory rhythms. The method is based on the spectral coherence, an index which quantifies the strength of the phase relationship between harmonics of the two signals occurring at the same frequency. The magnitude-squared coherence $MSC(f)$ between two signals $x(n)$ and $y(n)$ is defined as

$$MSC(f) = \frac{\left|S_{xy}(f)\right|^2}{S_{xx}(f)S_{yy}(f)} , \qquad (6.8)$$

where $S_{xy}(f)$ is the cross-spectrum of $x(n)$ and $y(n)$, and $S_{xx}(f)$ and $S_{yy}(f)$ are the individual power spectra of $x(n)$ and $y(n)$, respectively. The magnitude-squared coherence is normalized so that, at a given frequency, a value of 1 corresponds to complete coupling and 0 to absence of coupling. After the original paper [36], the spectral coherence was exploited to evaluate the degree of spatial linking in cardiac mapping applications [37]. One drawback of the coherence approach is that long sequences are required for robust spectral estimation, resulting in a method with low time resolution, thus limiting its applicability. For instance, when used to monitor the restoration of the normal sinus rhythm, the spectral coherence was not able to demonstrate an increase in AF organization the seconds before its termination [38]. A subsequent work resolved the problems related to time resolution by proposing a time–frequency coherence estimator with better time resolution [39]. The use of a time–frequency representation of the coherence allowed the authors to show transient electrical organization in the atria during AF, with coherence values increasing before either gradual or abrupt conversion to normal sinus rhythm.

Along with spectral estimation techniques, time domain measures of synchronization between atrial electrograms have been proposed to evaluate this specific aspect of AF organization. The first technique, proposed by Botteron and Smith [40], was based on crosscorrelation between closely spaced bipolar endocardial signals. The authors found that the atrial activation process during AF is spatially correlated with a degree of linking exponentially decaying with the distance between the two acquiring electrodes. Moreover, they proposed an activation space constant (ASC) to characterize, in a quantitative way, the decay of spatial organization with interelectrode distance [41, 40]. A major advantage of this approach is its relative computational simplicity. However, as for spectral coherence, some constraints related to temporal resolution still hold. Sih et al. [42] proposed an algorithm that measured synchronization between sites with the aim of obtaining a high temporal resolution. The method was based on quantifying linear relationships between short epochs (about 300 ms) of two different intra-atrial signals. It predicted the samples of one signal through linear adaptive filtering of the other. This approach was compared quantitatively with crosscorrelation and coherence measures, see Figure 6.1. The algorithm was adopted to quantify differences in organization between acute and chronic models of AF [43], and to provide evidence of heterogeneous remodeling in canine AF [44].

As coherence, crosscorrelation and adaptive filtering constitute linear approaches to the analysis of the synchronization between two fibrillatory electrograms, and therefore they cannot account for nonlinear coupling mechanisms. To fully account for the complexity of the dynamics underlying the coupling between the electrical activities of different sites during AF, measures of synchronization able to capture and quantify both linear and nonlinear interactions are required. This approach has been pursued by adapting to the analysis of endocardial signals some entropy-based methods proposed for the analysis of cardiovascular and cardiorespiratory interactions. Indeed, the studies of Censi et al. [45] and Mainardi et al. [9] estimated the degree of nonlinear coupling between pairs of bipolar endocardial electrograms acquired by decapolar catheters by performing specific multivariate embedding procedures. In particular, Censi et al. [45] generalized the concepts of correlation dimension and entropy defined in (6.6) and in (6.7), obtaining complexity and predictability indices of the nonlinear coupling between two atrial recordings. Mainardi et al. [9] estimated spatial organization in the atria by means of three indices assessing the coupling level between electrograms and evaluated the capability of the proposed indices to capture the fine changes in the dynamics of atrial signals induced by the injection of a sympathomimetic drug [46]. The indices included a crosscorrelation (CC) index, a nonlinear association (NLA) index, and a synchronization (S) index. From a methodological point of view, the three metrics provide information and quantify the coupling strength from different perspectives. The CC index accounts for the second-order moment of the signals, while neglecting the information contained in higher-order moments or in the nonlinear mechanisms. Conversely, these mechanisms are considered by both NLA and S. Briefly, the NLA can be obtained by computing the following ratio:

$$NLA = 1 - \frac{1}{\sigma_x^2(N-1)} \sum_{n=1}^{N} (x(n) - f(y(n)))^2, \qquad (6.9)$$

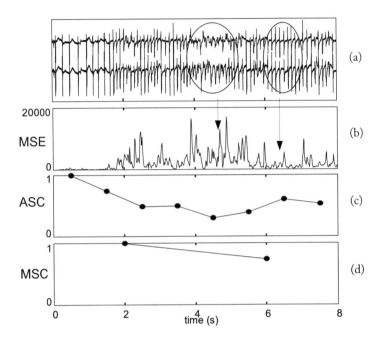

Figure 6.1: Comparison of organization measures between two epicardial electrograms (a) obtained over time during spontaneous conversion of atrial tachycardia to AF. The considered synchronization measures are (b) the mean-square error (MSE) resulting from linear prediction, (c) the activation space constant (ASC) based on crosscorrelation analysis, and (d) the magnitude squared coherence (MSC). All the methods reflect the decrease of organization moving from atrial tachycardia to AF, as demonstrated by the increase of MSE and the decrease of ASC and MSC after second 2. Moreover, MSE is shown to have a higher temporal resolution than ASC and MSC, and is capable of distinguishing the instances of lower and higher synchrony shown by the left and right circled areas, respectively. (Adapted from [42].)

where N is the series length, σ_x^2 is the variance of $x(n)$ and $f(y(n))$ is an interpolating function which represents an approximation of the curve describing the predicted value of $x(n)$ given $y(n)$, obtained as connection of linear segments [47]. Consequently, NLA expresses the variance reduction of $x(n)$ obtained by predicting the $x(n)$ values on the basis of $y(n)$: the better the prediction, the lower the unexplained variance and the higher the NLA index.

The degree of coupling between the two processes can also be obtained by means of the mutual-conditional entropy (MCE). Given a mixed-pattern $\mathbf{u}_{y/x}^L(m) = \begin{bmatrix} y(m) & \mathbf{x}^{L-1}(m) \end{bmatrix} = \begin{bmatrix} y(m) & x(m) & x(m-1) & \cdots & x(m-L+2) \end{bmatrix}$ of length L, its probability density function

$p(\mathbf{u}_{y/x}^{L}(m))$ may be estimated by

$$p(\mathbf{u}_{y/x}^{L}(m)) = \frac{1}{N - L + 1} \sum_{n=L}^{N} \Theta(\epsilon - \left\| \mathbf{u}_{y/x}^{L}(n) - \mathbf{u}_{y/x}^{L}(m) \right\|) \, . \tag{6.10}$$

The MCE is defined as [48]

$$MCE_{y/x}(L) = - \sum_{m} p(\mathbf{u}_{y/x}^{L}(m)) \log(\mathbf{u}_{y/x}^{L}(m)) - E(\mathbf{x}^{L-1}) \, , \tag{6.11}$$

where the summation is extended to all different mixed patterns. $MCE_{y/x}(L)$ measures the amount of information carried by the most recent sample of $y(n)$ when $L - 1$ samples of $x(n)$ are known and it reaches zero when a sufficient number of samples of $x(n)$ allows the complete prediction of $y(n)$. The index $MCE_{x/y}(L)$ suffers from the same limitation as CE_x (defined in Chapter 5): it decreases when L increases, independently of the type of coupling between $x(n)$ and $y(n)$, as an effect of the shortness of $x(n)$ and $y(n)$. Therefore, when short data segments are analyzed, a corrected $MCE_{y/x}(L)$ has to be utilized [48]:

$$MCCE_{y/x}(L) = MCE_{y/x}(L) - \text{perc}_{y/x}(L) \cdot E(\mathbf{y}^1) \, , \tag{6.12}$$

where $\text{perc}_{y/x}(L)$ is the percentage of mixed patterns found only once and $E(\mathbf{y}^1)$ is the Shannon entropy of the process. Taking into account that no a priori knowledge about the causal relationships between $x(n)$ and $y(n)$ is usually given (i.e., it is unknown if $y(n)$ affects $x(n)$ or vice versa), the synchronization index S is defined as

$$S = 1 - \min(U_{x,y}(L)) = 1 - \min \left(\frac{MCCE_{x/y}(L)}{E(\mathbf{x}^1)}, \frac{MCCE_{y/x}(L)}{E(\mathbf{y}^1)} \right) \, , \tag{6.13}$$

where $U_{x,y}(L)$ is known as the uncoupling function [48]. In (6.13), $MCCE_{y/x}(L)$ is normalized to avoid bias related to the shape of the probability distribution of the signals. The S index is sensitive to various signal coupling mechanisms (linear or not), including $1/N$ sub-harmonic links between signals [48], with superior performance with respect to CC as evidenced in many biosignal applications [9]. A comparative analysis of NLA and S shows that only conditional probability between samples are computed in the NLA estimate. Conversely, the conditional probabilities of patterns are also involved in the definition of S. Therefore, S is less sensitive to spurious coupling mechanisms and should be more reliable when the existence of coupling has to be excluded (Figure 6.2).

Coupling between atrial electrograms can also be assessed by quantifying the temporal synchronism between activation times in two sites. In this context, researchers focused their attention to either the atrial periods or the activation time sequences. Censi et al. [49] exploited recurrence plots to show that a certain degree of organization during AF can be detected as spatiotemporal recurrent patterns of the coupling between the atrial depolarization periods at two atrial sites. They demonstrated a deterministic mechanism underlying the apparently random activation processes

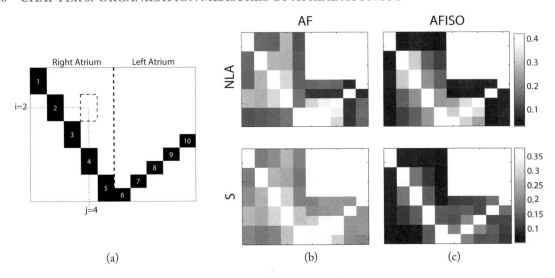

Figure 6.2: (a) Representation of electrode positions in the atria. A simplified representation of the atria with the position of the 10 electrodes: the first 5 electrodes map the whole right atrium, whereas the remaining 5 electrodes correspond to the inferior part of the left atrium as the catheter enters the coronary sinus. The intersection between line i and column j (belonging to the same atrium) represents the value of coupling of atrial signals recorded by the electrodes pair (i, j). The maps of (b) NLA and S indices during AF, and (c) AF during isoproterenol infusion (AFISO) for one patient. When passing from AF to AFISO, both indices captured the reduction in atrial organization due to adrenergic activation.

during AF. Barbaro et al. [50] proposed a synchronization index based on the assumption that two atrial regions can be considered as synchronized when they are depolarized by the same propagating wavefront. The proposed method was able to discriminate AF from regular atrial rhythms and to distinguish different degrees of complexity among AF episodes. The method was based on the analysis of the propagation delays between two atrial sites that were defined as the temporal interval between consecutive atrial activations at the two sites. A similar approach was proposed by Mase et al. [51] who characterized the synchronization between two atrial signals through a measure of the properties of the time delay distribution by the Shannon entropy. Specifically, the values of the propagation delay were quantized into M bins, and the entropy of their distribution was estimated as

$$H = -\sum_{i=1}^{m} p(i) \ln p(i) + \frac{m-1}{2N} , \tag{6.14}$$

where $m \leq M$ is the number of bins with nonzero probability, and $p(i)$ is the maximum likelihood estimate of the probability of having a propagation delay falling within the ith bin. The probability is computed as the ratio between the number of delays found in the ith bin and the total number of delays, denoted N. In (6.14), the first term is the classical definition of Shannon entropy, whereas the second term is introduced to reduce the systematic underestimation of Shannon entropy due to the approximation of the probabilities with the corresponding sample frequency. The synchronization between the two signals was then defined as

$$S_y = 1 - \frac{H}{\ln N} .$$
(6.15)

The resulting index S_y ranges from 0, when all N delays are in different bins (i.e., $H = \ln N$), to 1, when all delays are in the same bin (i.e., $H = 0$). The proposed synchronization index was validated with a computer model of atrial arrhythmias, and then applied to endocardial signals in AF patients [51]. It was shown to discriminate among different AF types and to elicit spatial heterogeneities in the synchronization between different atrial sites. Moreover, a comparison of the real data with simulation results linked the different shapes of the time delay distribution, and thus the S_y index, to different underlying electrophysiological propagation patterns. With respect to the methods analyzing atrial signals, approaches like this have the advantage of not being affected by the parameters which characterize the signal preprocessing, e.g., undersampling and filtering. On the other hand, they are highly dependent on the correct estimation of the atrial activation times, as explained in Section 6.3.1.

6.3.4   MORPHOLOGY-BASED ANALYSIS

The vast majority of the methods quantifying AF organization do not perform direct analysis of the morphologic features of the atrial signal. Most methods require significant manipulation of signal morphology to be effective; others, e.g., those performing pure analysis of event timing, do not consider morphology at all. Nevertheless, the morphology of the electrogram during activation of the atrial tissue contains relevant information about the process underlying the electrical propagation, as it basically reflects the interaction among the wavelets that sustain the arrhythmia. During AF, factors like refractoriness, conduction velocity, and atrial tissue heterogeneity bring about different types of electrical activation, such as slow conduction, wave collision, and conduction blocks [2]. These different activation mechanisms determine in turn the variations in morphology observed during different phases of the arrhythmia. For this reason, measures of organization which directly focus on the analysis of morphology of the atrial activation waves can be of interest for supporting both the investigation of the electrophysiological mechanisms underlying AF and its clinical treatment.

The historical attempt to describe morphology in endocardial recordings during AF was performed by Wells et al. [27] who provided qualitative criteria for classifying different episodes of the arrhythmia through visual electrogram scoring. From right atrial bipolar electrograms recorded after open-heart surgery, they classified AF recordings into four classes based on the discreteness of the electrograms, the variability in morphology, and the stability of the baseline. From then on, the assessment of the organization of bipolar recordings has mostly been achieved by visual scoring by expert cardiologists. The manual classification of atrial electrograms performed in accordance with Wells' criteria is indeed widely used to give morphologic characterization of AF, and has been related to the anatomic location and to the results of atrial pacing and catheter ablation [52, 53, 54]. In fact, it has been firmly established that, during AF, different regions of the atria exhibit different patterns of electrical activation [55, 56]. However, despite its widespread diffusion, Wells' approach is evidently limited as it is time consuming and can lead to results which are neither objective nor reproducible. To overcome these limitations, more quantitative and objective measures based on Wells' criteria were proposed, exploiting different signal processing approaches. In particular, Barbaro et al. [26] successfully developed an automatic classification scheme, including several time domain and frequency domain parameters, to typify AF signals according to Wells' criteria.

The classification of AF episodes into a limited number of types with increasing degrees of complexity is the most immediate approach to perform morphology-based analysis of organization. However, a more accurate evaluation of the features of the atrial signals may allow going beyond the determination of predefined classes of organization. In this context, Faes et al. [60, 57, 58] devised an approach for the evaluation of AF organization based on direct analysis of the variations in morphology of the atrial activation waves. In this approach, morphologic analysis of the atrial activations is performed through the definition of local activation waves (LAWs). For an endocardial signal in which N atrial depolarizations are detected, the LAWs $\mathbf{x}_i$, $i = 1, \ldots, N$, are defined as signal windows containing the whole atrial depolarization and centered at the local activation time (t_{x_i} in Figure 6.3). To prevent misleading variations in amplitude due to variable electrode–tissue contact, and to minimize the effects of bipolar electrodes orientation relative to the direction of wavefront propagation, the normalized LAWs $\mathbf{x}_i / \|\mathbf{x}_i\|$ were computed and used.

The categorization of AF episodes according to Wells' criteria was performed through principal component analysis (PCA) and cluster analysis [60], confirming the suitability of automatic and semi-automatic indices for online classification of AF episodes. With PCA, the large variety of morphologies corresponding to the activation waves recognized in a single endocardial signal during AF was reduced to a small subset of independent LAWs (the principal components), that were representative of the largest part of the original morphologic variability. A subsequent manual classification of the independent LAWs allowed semi-automatic classification of AF, with a dramatic reduction of the analysis time. Cluster analysis exploited an agglomerative approach to quantify the tendency of the LAWs to be grouped on the basis of their dissimilarity. To this end, the distance between two normalized LAWs, denoted $\mathbf{x}_i$ and $\mathbf{x}_j$, was first computed through the standard metric of the unitary sphere [60],

$$d(\mathbf{x}_i, \mathbf{x}_j) = \arccos(\mathbf{x}_i^T \mathbf{x}_j) \,. \tag{6.16}$$

Cluster analysis was performed by an iterative algorithm that started with N clusters, one for each LAW, and progressively joined the two clusters with maximal similarity until all LAWs were grouped into a single cluster. A clustering index was then defined as the average dissimilarity measured during the iteration of the algorithm. The index provided an automatic separation of different AF types on the basis of the morphologic similarity of the LAWs [60].

A method for quantifying the organization of single atrial electrograms was defined on the basis of a similarity analysis of the atrial activation waves [57]. The method compares the morpholo-

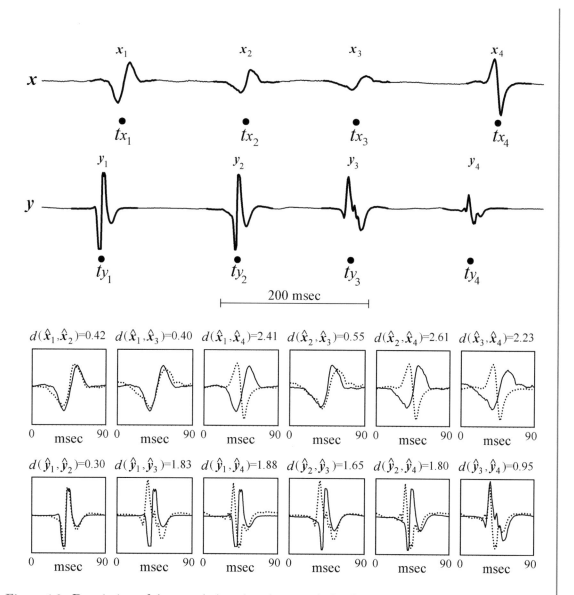

Figure 6.3: Description of the morphology-based approach for the evaluation of organization during AF [57, 58, 59]. From the bipolar endocardial signals shown in the upper traces, the local activation waves (LAWs) x_i and y_i are extracted as segments centered in the local activation times tx_i and ty_i. The normalized LAWs $\hat{x}_i$ and $\hat{y}_i$ are then compared pair-by-pair and their distance d is computed (lower boxes). (Reprinted from [58] with permission.)

gies of all possible pairs of LAWs. It yields a similarity index defined as the probability of finding similar LAWs within an endocardial signal:

$$\rho_x = \frac{2}{N(N-1)} \sum_{i=1}^{N} \sum_{j=i+1}^{N} \Theta(r - d(\mathbf{x}_i, \mathbf{x}_j)), \qquad (6.17)$$

where N is the number of LAWs, Θ the Heaviside function, and r a threshold below which two analyzed LAWs are considered similar; similarity between two normalized LAWs was assessed through the distance d defined in (6.16).

This formulation was extended to the study of coupling between two atrial signals, by the definition of an index incorporating information related both to morphologic and temporal information about the electrical activations at the two considered sites [58, 59]. After pairing the LAWs detected in two atrial signals $\mathbf{x}$ and $\mathbf{y}$, the probability of finding similar LAWs simultaneously in the two signals is estimated as:

$$\rho_{xy} = \frac{2}{N(N-1)} \sum_{i=1}^{N} \sum_{j=i+1}^{N} \Theta(r - \max(d(\mathbf{x}_i, \mathbf{x}_j), d(\mathbf{y}_i, \mathbf{y}_j))), \qquad (6.18)$$

where $\mathbf{x}_i$ and $\mathbf{y}_i$, $i = 1, \ldots, N$, are the paired normalized LAWs of the two signals (Figure 6.3). The coupling index associated with the two signals is then calculated as:

$$\chi_{xy} = \alpha \frac{2\rho_{xy}}{\rho_x + \rho_y}, \qquad (6.19)$$

where α is a temporal corrective term, accounting for the dispersion of the propagation delays between the two sites which are defined as the difference between the activation times marking the paired LAWs in the two signals [58].

The above-mentioned algorithms share the same concept to measure AF organization, i.e., they compare the morphologic features of the LAWs detected from endocardial signals to explore a specific aspect of organization. The rationale behind these algorithms is that an organized activation pattern produces repetitive and similar LAWs. Hence, the characterization of the changes occurring over time in waveform morphology represents the unifying framework on which the analysis performed by this approach is based. From this point of view, the quantification of regularity and coupling was integrated by the authors into a unified approach [58] in which the concept of morphologic similarity was stressed and practical applications of the indices were prospected. Specifically, the multi-site evaluation of morphology-based indices of regularity and coupling was shown to be useful to investigate the spatial distribution of AF organization, that may reflect critical activation patterns for the maintenance of the arrhythmia. In particular, in the study by Ravelli et al. [61], maps of wave similarity were constructed during paroxysmal and chronic AF by extending to the whole right atrial surface the calculation of the index in (6.17) after acquisition of simultaneous recording through multipolar basket catheter; see Figure 6.4. Wave similarity mapping of the right atrium

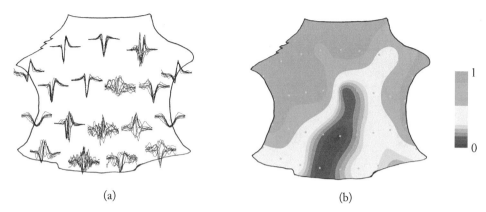

(a)   (b)

Figure 6.4: Construction of wave similarity maps during paroxysmal AF. (a) The superposition of local activation waves shows the spatial distribution of waveform complexity, while (b) the corresponding values of the similarity index of (6.17), interpolated over the whole right atrial surface and color coded, indicate the presence of areas with different degree of organization. (Adapted from [61].)

evidenced the anatomic anchoring of the electrical pattern in paroxysmal AF, with high regularity in the anterolateral wall and low regularity in the posteroseptal region, and a reduction of the islands of organization in chronic AF, supporting the new ablative strategies for substrate modification [62]. Moreover, capturing the dynamical evolution of AF organization should be useful both to investigate the mechanisms underlying the development of the arrhythmia [63], and to optimize pacing protocols aimed to restore the normal rhythm in AF patients [64, 65]. Ravelli et al. [66] reported for the first time the quantitative evolution of wave similarity during the first minutes of AF, performing a beat-to-beat evaluation of the similarity index in (6.17); see Figure 6.5. The study showed that high levels of organization exist at the onset of paroxysmal AF, with progressive deterioration of organization, characterized by a decrease of the similarity index, within the first three minutes after AF onset. As antitachycardia pacing success rate depends on the organization state [65], this result suggests an early delivery of pacing treatment.

6.4   CONCLUSIONS

The present chapter documents the development of a large variety of methods evaluating the degree of organization of the electrical activity during AF through the analysis of endocardial or epicardial atrial signals. The reasons for this proliferation reside on the growing possibilities offered by the recent advances in the acquisition and processing of electrophysiological signals from the atrial chambers, and on the fact that an accurate extraction of information from the atrial signals may play an active role both in the investigation of the arrhythmia mechanisms and in the support of the new therapeutic approaches. Nonetheless, the existence of such a large body of work prompts for categorization of the methods measuring AF organization. This need comes also from the consideration that

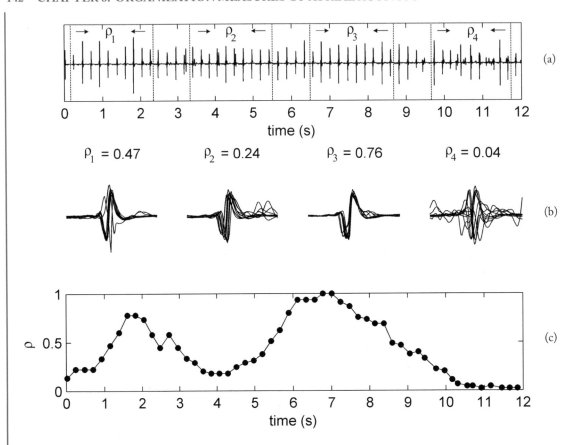

Figure 6.5: Time course of AF organization using the morphology-based approach. (a) The endocardial signal presents epochs with different degree of organization as shown in (b) the different superposition of the activation waves in the four windows, that correspond to different values of the similarity index in (6.18). (c) The index was able to capture the evolution of AF organization with finer temporal resolution as shown by the beat-to-beat evaluation. (Adapted from [66].)

the term "organization" is not univocally defined in AF analysis, and subjective interpretations of the underlying concept are commonly given when a new signal processing method is proposed. In this chapter, the methods quantifying AF organization were categorized by considering which aspects affecting AF are the most important for the methods themselves. This would facilitate the comparison of performance of the various methods: if indices are subdivided by the way they determine *how AF is organized*, it should be easier to evaluate and compare their ability in quantifying *how organized AF is*. Moreover, this philosophy in categorizing methods should facilitate the selection of the method that is most appropriate for a specific task, such as its application within a given therapeutic approach, pharmacological evaluation, mechanisms investigation, and so on.

In the proposed categorization, four different meanings of the concept of organization during AF are recognized, each one reflecting specific electrophysiological aspects and methodological peculiarities. The existence of specific and well-defined oscillations, documenting the presence of a periodic activity, is investigated by methods performing rhythm analysis; the dynamical complexity of the electrical activity of specific atrial regions, mostly related to the local wave fragmentation, is evaluated by regularity analysis; the spatial coordination between the activity of different atrial sites and thus the mechanisms of wave propagation, is considered by the synchronization analysis; the similarity over time of the activation wave morphologies, reflecting the stability of the underlying activation pattern, is the basic element by which morphologic analysis is performed. All these approaches have been followed to quantify the organization of atrial electrical activity during fibrillation, and have been successfully exploited in both basic research and clinical applications. Indeed, spatial and temporal measures of organization have been used for recognizing specific patterns of wavelet interaction and relating such patterns with functional and structural characteristics of the atria, as well as for advancing strategies for the treatment of AF such as catheter ablation and electrical cardioversion.

Bibliography

[1] J. Jalife, "Rotors and spiral waves in atrial fibrillation," *J. Cardiovasc. Electrophysiol.*, vol. 14, pp. 776–780, 2003. DOI: 10.1046/j.1540-8167.2003.03136.x

[2] K. T. Konings, J. L. Smeets, O. C. Penn, H. J. Wellens, and M. A. Allessie, "Configuration of unipolar atrial electrograms during electrically induced atrial fibrillation in humans," *Circulation*, vol. 95, pp. 1231–241, 1997.

[3] K. T. Konings, C. J. Kirchhof, J. R. Smeets, H. J. Wellens, O. C. Penn, and M. A. Allessie, "High-density mapping of electrically induced atrial fibrillation in humans," *Circulation*, vol. 89, pp. 1665–1680, 1994.

[4] H. J. Sih, "Measures of organization during atrial fibrillation," *Ann. Ist. Super. Sanita*, vol. 37, pp. 361–369, 2001.

[5] V. Barbaro, P. Bartolini, G. Calcagnini, and F. Censi, "Extraction of physiological and clinical information from intra-atrial electrograms during atrial fibrillation: review of methods," *Ann. Ist. Super. Sanita*, vol. 37, pp. 319–324, 2001.

[6] A. Capucci, M. Biffi, G. Boriani, F. Ravelli, G. Nollo, P. Sabbatani, C. Orsi, and B. Magnani, "Dynamic electrophysiological behavior of human atria during paroxysmal atrial fibrillation," *Circulation*, vol. 92, pp. 1193–1202, 1995.

[7] D. O'Donnell, S. S. Furniss, and J. P. Bourke, "Paroxysmal cycle length shortening in the pulmonary veins during atrial fibrillation correlates with arrhythmogenic triggering foci in sinus rhythm," *J. Cardiovasc. Electrophysiol.*, vol. 13, pp. 124–128, 2002. DOI: 10.1046/j.1540-8167.2002.00124.x

[8] P. Sanders, C. J. Nalliah, R. Dubois, Y. Takahashi, M. Hocini, M. Rotter, T. Rostock, F. Sacher, L. F. Hsu, A. Jonsson, M. D. O'Neill, P. Jais, and M. Haissaguerre, "Frequency mapping of the pulmonary veins in paroxysmal versus permanent atrial fibrillation," *J. Cardiovasc. Electrophysiol.*, vol. 17, pp. 965–972, 2006. DOI: 10.1111/j.1540-8167.2006.00546.x

[9] L. T. Mainardi, A. Porta, G. Calcagnini, P. Bartolini, A. Michelucci, and S. Cerutti, "Linear and nonlinear analysis of atrial signals and local activation period series during atrial fibrillation episodes," *Med. Biol. Eng. & Comput.*, vol. 39, pp. 249–254, 2001. DOI: 10.1007/BF02344809

[10] L. T. Mainardi, V. D. A. Corino, L. Lombardi, C. Tondo, M. Mantica, F. Lombardi, and S. Cerutti, "Assessment of the dynamics of atrial signals and local atrial period series during atrial fibrillation: effects of isoproterenol administration," *Biomedical Eng. OnLine*, vol. 3, p. 37, 2004. DOI: 10.1186/1475-925X-3-37

[11] C. F. Pieper and A. Pacifico, "Improved spatial resolution for cardiac mapping using current source density-based electrode arrays," *Pacing Clin. Electrophysiol.*, vol. 18, pp. 34–44, 1995. DOI: 10.1111/j.1540-8159.1995.tb02473.x

[12] G. Ndrepepa, E. B. Caref, H. Yin, N. E. Sherif, and M. Restivo, "Activation time determination by high-resolution unipolar and bipolar extracellular electrograms in the canine heart," *J. Cardiovasc. Electrophysiol.*, vol. 6, pp. 174–188, 1995. DOI: 10.1111/j.1540-8167.1995.tb00769.x

[13] R. Antolini, M. Kirchner, A. Mongera, M. Disertori, and F. Furlanello, "On-line interval measurement during invasive cardiac electrophysiologic testing," *Pacing Clin. Electrophysiol.*, vol. 11, pp. 33–46, 1988. DOI: 10.1111/j.1540-8159.1988.tb03928.x

[14] E. V. Simpson, R. E. Ideker, C. Cabo, S. Yabe, X. Zhou, S. B. Melnick, and W. M. Smith, "Evaluation of an automatic cardiac activation detector for bipolar electrograms," *Med. Biol. Eng. & Comput.*, vol. 31, pp. 118–128, 1993. DOI: 10.1007/BF02446669

[15] C. F. Pieper, R. Blue, and A. Pacifico, "Simultaneously collected monopolar and discrete bipolar electrograms: comparison of activation time detection algorithms," *Pacing Clin. Electrophysiol.*, vol. 16, pp. 426–433, 1993. DOI: 10.1111/j.1540-8159.1993.tb01605.x

[16] L. Sandrini, L. Faes, F. Ravelli, R. Antolini, and G. Nollo, "Morphology-based measurement of activation time in human atrial fibrillation," in *Proc. Comput. Cardiol.*, vol. 29, pp. 593–596, IEEE Press, 2002.

[17] A. C. Skanes, R. Mandapati, O. Berenfeld, J. M. Davidenko, and J. Jalife, "Spatiotemporal periodicity during atrial fibrillation in the isolated sheep heart," *Circulation*, vol. 98, pp. 1236–1248, 1998.

[18] M. Mansour, R. Mandapati, O. Berenfeld, J. Chen, F. H. Samie, and J. Jalife, "Left-to-right gradient of atrial frequencies during acute atrial fibrillation in the isolated sheep heart," *Circulation*, vol. 103, pp. 2631–2636, 2001.

[19] S. Lazar, S. Dixit, F. E. Marchlinski, D. J. Callans, and E. P. Gerstenfeld, "Presence of left-to-right atrial frequency gradient in paroxysmal but not persistent atrial fibrillation in humans," *Circulation*, vol. 110, pp. 3181–3186, 2004. DOI: 10.1161/01.CIR.0000147279.91094.5E

[20] S. Lazar, S. Dixit, D. J. Callans, D. Lin, F. E. Marchlinski, and E. P. Gerstenfeld, "Effect of pulmonary vein isolation on the left-to-right atrial dominant frequency gradient in human atrial fibrillation," *Heart Rhythm*, vol. 3, pp. 889–895, 2006. DOI: 10.1016/j.hrthm.2006.04.018

[21] T. H. Everett, L. C. Kok, R. H. Vaughn, J. R. Moorman, and D. E. Haines, "Frequency domain algorithm for quantifying atrial fibrillation organization to increase defibrillation efficacy," *IEEE Trans. Biomed. Eng.*, vol. 48, pp. 969–978, 2001. DOI: 10.1109/10.942586

[22] T. H. Everett, J. R. Moorman, L. C. Kok, J. G. Akar, and D. E. Haines, "Assessment of global atrial fibrillation organization to optimize timing of atrial defibrillation," *Circulation*, vol. 103, pp. 2857–2861, 2001.

[23] T. H. Everett, J. G. Akar, L. C. Kok, J. R. Moorman, and D. E. Haines, "Use of global atrial fibrillation organization to optimize the success of burst pace termination," *J. Am. Coll. Cardiol.*, vol. 40, pp. 1831–1840, 2002. DOI: 10.1016/S0735-1097(02)02476-2

[24] Y. Takahashi, P. Sanders, P. Jais, M. Hocini, R. Dubois, M. Rotter, T. Rostock, C. J. Nalliah, F. Sacher, J. Clementy, , and M. Haissaguerre, "Organization of frequency spectra of atrial fibrillation: relevance to radiofrequency catheter ablation," *J. Cardiovasc. Electrophysiol.*, vol. 17, pp. 382–388, 2006. DOI: 10.1111/j.1540-8167.2005.00414.x

[25] V. Barbaro, P. Bartolini, G. Calcagnini, F. Censi, S. Morelli, and A. Michelucci, "Mapping the organization of atrial fibrillation with basket catheters. part i: Validation of a real-time algorithm," *Pacing Clin. Electrophysiol.*, vol. 24, pp. 1082–1088, 2001. DOI: 10.1046/j.1460-9592.2001.01082.x

[26] V. Barbaro, P. Bartolini, G. Calcagnini, S. Morelli, A. Michelucci, and G. Gensini, "Automated classification of human atrial fibrillation from intraatrial electrograms," *Pacing Clin. Electrophysiol.*, vol. 23, pp. 192–202, 2000. DOI: 10.1111/j.1540-8159.2000.tb00800.x

[27] J. L. Wells, R. B. Karp, N. T. Kouchoukos, W. A. MacLean, T. N. James, and A. L. Waldo, "Characterization of atrial fibrillation in man: studies following open heart surgery," *Pacing Clin. Electrophysiol.*, vol. 1, pp. 426–438, 1978. DOI: 10.1111/j.1540-8159.1978.tb03504.x

[28] A. Michelucci, P. Bartolini, G. Calcagnini, F. Censi, A. Colella, S. Morelli, L. Padeletti, P. Pieragnoli, and V. Barbaro, "Mapping the organization of atrial fibrillation with basket catheters. part ii: Regional patterns in chronic patients," *Pacing Clin. Electrophysiol.*, vol. 24, pp. 1089–1096, 2001. DOI: 10.1046/j.1460-9592.2001.01089.x

[29] D. R. Chialvo, R. F. Gilmour, and J. Jalife, "Low dimensional chaos in cardiac tissue," *Nature*, vol. 343, pp. 653–657, 1990. DOI: 10.1038/343653a0

[30] D. T. Kaplan and R. J. Cohen, "Is fibrillation chaos?," *Circ. Res.*, vol. 67, pp. 886–892, 1990.

[31] B. P. T. Hoekstra, C. G. H. Diks, M. A. Allessie, and J. De Goede, "Nonlinear analysis of epicardial atrial electrograms of electrically induced atrial fibrillation in man," *J. Cardiovasc. Electrophysiol.*, vol. 6, pp. 419–440, 1995. DOI: 10.1111/j.1540-8167.1995.tb00416.x

[32] H. F. Pitschner, A. Berkovic, S. Grumbrecht, and J. Neuzner, "Multielectrode basket catheter mapping for human atrial fibrillation," *J. Cardiovasc. Electrophysiol.*, vol. 9, pp. S48–S56, 1998.

[33] A. Berkowitsch, A. Erdogan, J. Neuzner, and H. F. Pitschner, "New non linear approach to a quantitative assessment of atrial fibrillation mapping," *Med. Biol. Eng. & Comput.*, vol. 37, pp. 398–399, 1999.

[34] H. F. Pitschner and A. Berkowitsch, "Algorithmic complexity. a new approach of nonlinear algorithms for the analysis of atrial signals from multipolar basket catheter," *Ann. Ist. Super. Sanita*, vol. 37, pp. 409–418, 2001.

[35] L. T. Mainardi, A. Porta, G. Calcagnini, F. Censi, P. Bartolini, A. Michelucci, and S. Cerutti, "Discrimination of atrial rhythms by linear and nonlinear methods," *Ann. Ist. Super. Sanita*, vol. 37, pp. 335–340, 2001.

[36] K. M. Ropella, A. V. Sahakian, J. M. Baerman, and S. Swiryn, "The coherence spectrum. A quantitative discriminator of fibrillatory and nonfibrillatory cardiac rhythms," *Circulation*, vol. 80, pp. 112–119, 1989.

[37] H. J. Sih, A. V. Sahakian, C. E. Arentzen, and S. Swiryn, "A frequency domain analysis of spatial organization of epicardial maps," *IEEE Trans. Biomed. Eng.*, vol. 42, pp. 718–727, 1995. DOI: 10.1109/10.391158

[38] H. J. Sih, K. M. Ropella, S. Swiryn, E. P. Gerstenfeld, and A. V. Sahakian, "Observations from intraatrial recordings on the termination of electrically induced atrial fibrillation in humans," *Pacing Clin. Electrophysiol.*, vol. 17, pp. 1231–1242, 1994. DOI: 10.1111/j.1540-8159.1994.tb01490.x

[39] E. G. Lovett and K. M. Ropella, "Time–frequency coherence analysis of atrial fibrillation termination during procainamide administration," *Ann. Biomed. Eng*, vol. 25, pp. 975–984, 1997. DOI: 10.1007/BF02648123

[40] G. W. Botteron and J. M. Smith, "A technique for measurement of the extent of spatial organization of atrial activation during atrial fibrillation in the intact human heart," *IEEE Trans. Biomed. Eng.*, vol. 42, pp. 579–586, 1995. DOI: 10.1109/10.387197

[41] G. W. Botteron and J. M. Smith, "Quantitative assessment of the spatial organization of atrial fibrillation in the intact human heart," *Circulation*, vol. 93, pp. 513–518, 1996.

[42] H. J. Sih, D. P. Zipes, E. J. Berbari, and J. E. Olgin, "A high-temporal resolution algorithm for quantifying organization during atrial fibrillation," *IEEE Trans. Biomed. Eng.*, vol. 46, pp. 440–450, 1999. DOI: 10.1109/10.752941

[43] H. J. Sih, D. P. Zipes, E. J. Berbari, D. E. Adams, and J. E. Olgin, "Differences in organization between acute and chronic atrial fibrillation in dogs," *J. Am. Coll. Cardiol.*, vol. 36, pp. 924–931, 2000. DOI: 10.1016/S0735-1097(00)00788-9

[44] H. J. Sih, "Evidence of heterogeneous remodeling in canine atrial fibrillation.," *J. Electrocardiol.*, vol. 33 suppl, pp. 141–145, 2000. DOI: 10.1054/jelc.2000.20306

[45] F. Censi, V. Barbaro, P. Bartolini, G. Calcagnini, A. Michelucci, and S. Cerutti, "Nonlinear coupling of atrial activation processes during atrial fibrillation in humans," *Biol. Cybern.*, vol. 85, pp. 195–201, 2001. DOI: 10.1007/s004220100252

[46] L. T. Mainardi, V. D. Corino, L. Lombardi, C. Tondo, M. Mantica, F. Lombardi, and S. Cerutti, "Linear and nonlinear coupling between atrial signals. Three methods for the analysis of the relationships among atrial electrical activities in different sites," *IEEE Eng. Med. Biol. Mag.*, vol. 25, pp. 63–70, 2006. DOI: 10.1109/EMB-M.2006.250509

[47] J. P. Pijn and F. L. da Silva, "Propagation of electrical activity: nonlinear associations and time delays between EEG signals," in *Basic Mechanisms of the EEG* (S. Zschocke and E. J. Speckmann, eds.), pp. 41–59, Boston: Birkhauser, 1993.

[48] A. Porta, G. Baselli, F. Lombardi, N. Montano, A. Malliani, and S. Cerutti, "Conditional entropy approach for the evaluation of the coupling strength," *Biol. Cybern.*, vol. 81, pp. 119–129, 1999. DOI: 10.1007/s004220050549

[49] F. Censi, V. Barbaro, P. Bartolini, G. Calcagnini, A. Michelucci, G. F. Gensini, and S. Cerutti, "Recurrent patterns of atrial depolarization during atrial fibrillation assessed by recurrence plot quantification," *Ann. Biomed. Eng.*, vol. 28, pp. 61–70, 2000. DOI: 10.1114/1.248

[50] V. Barbaro, P. Bartolini, G. Calcagnini, F. Censi, and A. Michelucci, "Measure of synchronisation of right atrial depolarisation wavefronts during atrial fibrillation," *Med. Biol. Eng. & Comput.*, vol. 40, pp. 56–62, 2002. DOI: 10.1007/BF02347696

[51] M. Mase, L. Faes, R. Antolini, M. Scaglione, and F. Ravelli, "Quantification of synchronization during atrial fibrillation by Shannon entropy: validation in patients and computer model of atrial arrhythmias," *Physiol Meas.*, vol. 26, pp. 911–923, 2005. DOI: 10.1088/0967-3334/26/6/003

[52] H. Li, J. Hare, K. Mughal, D. Krum, M. Biehl, S. Deshpande, A. Dhala, Z. Blanck, J. Sra, M. Jazayeri, and M. Akhtar, "Distribution of atrial electrogram types during atrial fibrillation: effect of rapid atrial pacing and intercaval junction ablation," *J. Am. Coll. Cardiol.*, vol. 27, pp. 1713–1721, 1996. DOI: 10.1016/0735-1097(96)00068-X

[53] F. Gaita, R. Riccardi, L. Calo, M. Scaglione, L. Garberoglio, R. Antolini, M. Kirchner, F. Lamberti, and E. Richiardi, "Atrial mapping and radiofrequency catheter ablation in patients with idiopathic atrial fibrillation. electrophysiological findings and ablation results," *Circulation*, vol. 97, pp. 2136–2145, 1998.

[54] M. Haissaguerre, L. Gencel, B. Fischer, P. L. Metayer, F. Poquet, F. I. Marcus, and J. Clementy, "Successful catheter ablation of atrial fibrillation," *J. Cardiovasc. Electrophysiol.*, vol. 5, pp. 1045–1052, 1994. DOI: 10.1111/j.1540-8167.1994.tb01146.x

[55] P. Jais, M. Haissaguerre, D. C. Shah, S. Chouairi, and J. Clementy, "Regional disparities of endocardial atrial activation in paroxysmal atrial fibrillation," *Pacing Clin. Electrophysiol.*, vol. 19, pp. 1998–2003, 1996. DOI: 10.1111/j.1540-8159.1996.tb03269.x

[56] F. Gaita, L. Calo, R. Riccardi, L. Garberoglio, M. Scaglione, G. Licciardello, L. Coda, P. D. Donna, M. Bocchiardo, D. Caponi, R. Antolini, F. Orzan, and G. Trevi, "Different patterns of atrial activation in idiopathic atrial fibrillation: simultaneous multisite atrial mapping in patients with paroxysmal and chronic atrial fibrillation," *J. Am. Coll. Cardiol.*, vol. 37, pp. 534–541, 2001. DOI: 10.1016/S0735-1097(00)01120-7

[57] L. Faes, G. Nollo, R. Antolini, F. Gaita, and F. Ravelli, "A method for quantifying atrial fibrillation organization based on wave morphology similarity," *IEEE Trans. Biomed. Eng.*, vol. 49, pp. 1504–1513, 2002. DOI: 10.1109/TBME.2002.805472

[58] L. Faes and F. Ravelli, "A morphology-based approach to the evaluation of atrial fibrillation organization," *IEEE Eng. Med. Biol. Mag.*, vol. 26, pp. 59–67, 2007. DOI: 10.1109/MEMB.2007.384097

[59] L. Faes, L. Sandrini, F. Ravelli, R. Antolini, and G. Nollo, "Quantitative assessment of regularity and synchronization of intracardiac recordings during human atrial fibrillation," in *Proc. Comput. Cardiol.*, vol. 29, pp. 597–600, IEEE Press, 2002.

[60] L. Faes, G. Nollo, M. Kirchner, E. Olivetti, F. Gaita, R. Riccardi, and R. Antolini, "Principal component analysis and cluster analysis for measuring the local organisation of human atrial fibrillation," *Med. Biol. Eng. & Comput.*, vol. 39, pp. 656–663, 2001. DOI: 10.1007/BF02345438

[61] F. Ravelli, L. Faes, L. Sandrini, F. Gaita, R. Antolini, M. Scaglione, and G. Nollo, "Wave similarity mapping shows the spatiotemporal distribution of fibrillatory wave complexity in the human right atrium during paroxysmal and chronic atrial fibrillation," *J. Cardiovasc. Electrophysiol.*, vol. 16, pp. 1071–1076, 2005. DOI: 10.1111/j.1540-8167.2005.50008.x

[62] K. Nademanee, J. McKenzie, E. Kosar, M. Schwab, B. Sunsaneewitayakul, T. Vasavakul, C. Khunnawat, and T. Ngarmukos, "A new approach for catheter ablation of atrial fibrillation: mapping of the electrophysiologic substrate," *J. Am. Coll. Cardiol.*, vol. 43, pp. 2044–2053, 2004. DOI: 10.1016/j.jacc.2003.12.054

[63] M. Allessie, J. Ausma, and U. Schotten, "Electrical, contractile and structural remodeling during atrial fibrillation," *Cardiovasc. Res.*, vol. 54, pp. 230–246, 2002. DOI: 10.1016/S0008-6363(02)00258-4

[64] A. Capucci, F. Ravelli, G. Nollo, A. S. Montenero, M. Biffi, and G. Q. Villani, "Capture window in human atrial fibrillation: evidence of an excitable gap," *J. Cardiovasc. Electrophysiol.*, vol. 10, pp. 319–327, 1999. DOI: 10.1111/j.1540-8167.1999.tb00678.x

[65] C. W. Israel, J. R. Ehrlich, G. Gronefeld, A. Klesius, T. Lawo, B. Lemke, and S. H. Hohnloser, "Prevalence, characteristics and clinical implications of regular atrial tachyarrhythmias in patients with atrial fibrillation: insights from a study using a new implantable device," *J. Am. Coll. Cardiol.*, vol. 38, pp. 355–363, 2001. DOI: 10.1016/S0735-1097(01)01351-1

[66] F. Ravelli, M. Mase, M. D. Greco, L. Faes, and M. Disertori, "Deterioration of organization in the first minutes of atrial fibrillation: a beat-to-beat analysis of cycle length and wave similarity," *J. Cardiovasc. Electrophysiol.*, vol. 18, pp. 60–65, 2007. DOI: 10.1111/j.1540-8167.2006.00620.x

CHAPTER 7

Modeling Atrial Fibrillation: From Myocardial Cells to ECG

Adriaan van Oosterom and Vincent Jacquemet

7.1 INTRODUCTION

The standard 12-lead ECG is the most commonly used noninvasive tool for diagnosing cardiac electrically manifest abnormalities like those of arrhythmias. It presents simultaneously recorded electrical signals, derived from nine electrodes placed on the thorax and the extremities. The waveforms of the signals are lead-specific. An example of an ECG, recorded during a normal heart rhythm, is shown in the upper trace of Figure 7.1. The PQRSTU nomenclature indicated was introduced by Einthoven [1]. In contrast, the middle trace of Figure 7.1 shows the typical waveform observed in lead V1 in a patient during atrial fibrillation (AF). Here the atrial activity is continuous, producing signals with irregular waveforms superimposed on the ventricular signals. The ventricular heart rate is more irregular than during sinus rhythm. However, it is less irregular than the fibrillating atrium owing to the fact that the electric trigger from atrial to ventricular tissue passes through, and is regulated by the specialized cells of the atrioventricular (AV) node. The atrial signals are clearly much smaller, roughly 10 times, than those related to the ventricles (QRST complexes), both during normal rhythm (P waves) and arrhythmias like AF. As a consequence, the early diagnostic applications of the P wave were necessarily limited, and the use of the ECG during AF was mainly restricted to diagnosing the presence of AF as such.

Besides using the ECG signals, the electrical activity of the atria is recorded by passing catheter-guided electrodes through one of the major veins into the atria. The analysis of the recorded signals (endocardial electrograms), with their far greater temporal detail, is used in diagnostic procedures as well as for guiding therapeutic interventions [3]. Advanced mapping procedures have been developed for scanning the electrical activity on the exposed atrial surface [4] or endocardium [5]. Recent developments include the application of such methods in devices such as CARTO (Biosense Webster) and EnSite (St Jude Medical). In the CARTO method, subsequent single-lead recordings of the potential are taken at different locations on the endocardium, while at the same time building up a geometric representation of the atrial surface from the electrode locations, which sample 3D space. The EnSite method involves the simultaneous sampling of the potentials at a regular grid

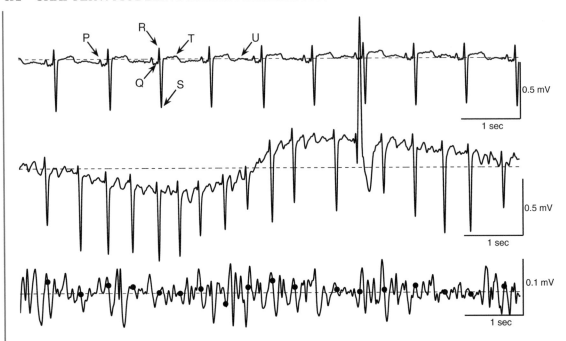

Figure 7.1: Waveforms of standard lead V1. Upper trace: healthy subject during sinus rhythm. Middle trace: clinical recording of AF patient; note the wandering baseline and the presence of an ectopic beat. Such beats frequently occur during AF. Lower trace: signal of the middle trace after baseline correction and suppression of the ventricular (QRST) involvement. Note its different scaling. The dots mark the timing of the S waves shown in the middle trace. (Adapted from [2].)

placed within the atria. The endocardial potential field and signals are derived from a computational inverse procedure. At present the CARTO method does not provide simultaneous recordings and, hence, the application to arrhythmia is limited. The inverse procedure involved in the processing of the simultaneous recordings used by the EnSite device has only limited accuracy. Such methods have led to the identification of different types of foci that may trigger AF, the description of their specific dynamics, and their location.

In contrast to a catheterization procedure, the recording of conventional ECGs from the thorax is noninvasive. The ECG provides a global, overall impression of the atrial activity, an impression that catheterization provides only through an elaborate, time consuming procedure. However, when relying on body surface potentials, an unfortunate consequence of the spreading out of electrical current is that the electrical image that appears on the body surface is blurred: spatial and temporal definition is less than that of potentials observed on the atria. Their amplitudes are about 10 times smaller than those recorded by means of catheters.

Current basic research is aimed at linking the epi- or endocardial signals with different etiologies of the disease. In view of an ultimate application of the insight gained in clinical diagnostic procedures, it is of interest to study how much of the spatiotemporal complexity observed at the atria may be retained in signals observed on the body surface.

The development of signal processing tools aimed at isolating the contribution of the atria to the ECG signals during AF (Chapter 3) has opened up a clear view on these signals (lower trace of Figure 7.1). This opens the way to the full extraction of the spatiotemporal information of AF signals and has strengthened the motivation for studying the diagnostic potential of body surface signals.

This chapter describes a biophysical model of the electrical sources of the atria during AF, derived from membrane kinetics. In addition, it is shown how the transfer between electrical sources and resulting potentials may be computed using a volume conductor. The combination of both models can be used to study the amplitudes and waveforms of signals as observed invasively as well as on the thorax. The model based approach permits the study of atrial signals on the body surface that are completely free of contamination by ventricular electrical activity. Examples of some early applications of this model based approach can be found in [6].

7.2   GENESIS OF ELECTROCARDIOGRAPHIC POTENTIALS: THE FORWARD PROBLEM

When interpreting the electrical signals arising from the heart's activity two separate factors are invariably involved. The first is the specification of the (active) electrical current sources. The second, equally important factor is a description of the effects of the passive electrical properties of the medium surrounding the heart: the body tissues. The currents generated by the heart's active properties pass through this medium, where they set up the potential differences observed between the two terminals of any lead system used.

In their diagnostic application of observed ECG waveforms, most cardiologists (and electrophysiologists) are usually not aware of these separate factors and they perform a mental signal analysis on the observed signals based on training and past experience. However, when trying to stretch the diagnostic performance of ECG-based techniques both factors need to be considered.

The computation of the electrical potential field generated by cardiac electrical sources is generally known as the *forward problem of electrocardiography*. In solving this problem, the two factors mentioned before are handled by formulating a model of the electrical cardiac sources, the *source model*, as well as of the conductive tissues surrounding these sources, the *volume conduction model*. Due to the complexity of living tissue, morphology as well as function-wise, a full description of these factors is beyond reach. Instead, different models have been postulated for handling both factors, with their complexity tuned to the particular application in hand. The models discussed in this chapter aim at providing a description of the genesis of atrial signals that does justice to the physical and electrophysiological complexity involved. A more complete treatment of the forward problem can be found in several excellent textbooks, such as [7, 8, 9]. The models used in our current work as well as their background are summarized in the following two sections.

7.3 SOURCE MODELS

The single current dipole is probably the best known model of the cardiac electrical generator. In Einthoven's papers, as well as in his correspondence with Lewis and Wilson [10], it can be seen as an arrow drawn on a plane, representing the electromotive force of the heart. However, the dipole cannot be interpreted directly in terms of the underlying electrophysiology, and certainly not for modeling, in a forward sense, the electrophysiological complexity of AF. It is for this application that the source model described below holds more promise.

7.3.1 ELECTROPHYSIOLOGICAL BACKGROUND

The contractile elements of the atria are located inside the myocytes, a structure of densely packed cells that form a muscle layer. The overall geometry of the atrial model used is shown in Figure 7.2. The hull shown, containing the myocytes, forms a *closed* surface in the sense that any of its interior points may be connected to any other interior point while following a route that lies entirely within the hull.

The activation and deactivation process of the contractile elements is controlled by calcium ions (Ca^{2+}), the intracellular concentration of which is gated by the instantaneous potential difference across the cell membrane, the transmembrane potential (TMP): $V_m(t)$, the potential of the intracellular region minus that of the extracellular medium just outside the cell membrane. This potential difference depends on the concentrations of various other ions, in particular the sodium (Na^+) and potassium (K^+) ions. These "gatekeepers" of the associated fluxes of ions across the membrane are localized in the cell membrane. They are the so-called ion channels, for which several different, ion-specific variants have been identified. The entire process of activation and recovery arises from the complex, dynamic interplay between these ions, fed by the underlying biochemical processes that supply the required contractile energy.

At the end of the atrial diastolic interval, all TMP levels are close to their resting state: a polarized state with an interior potential of about -80 mV. Following an initiating stimulus, the TMP rises rapidly to a level of about $+5$ mV, a polarity reversal generally called *depolarization*, facilitating the activation of the contractile elements. Once a sufficiently large region is active, the polarized myocytes at the border of the activated region are depolarized by an electrical current flow from the activated region. In turn, their activation facilitates that of any neighbors still at rest. In this manner, a wave-like activation process is set up. Following activation, the myocytes gradually return to their resting state. During the first part of this recovery phase, the so-called refractory period, reactivation is impossible.

In a normal heart, the initiation of the activity stems from a structure of specialized cells, the so-called sino-atrial node (SAN), which is located in the right atrium close to the superior vena cava (Figure 7.2). From there the activation propagates in a regular fashion over the atrial tissue at

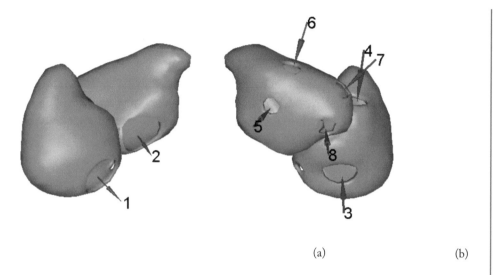

(a)   (b)

Figure 7.2: Atrial geometry; wall thickness approximately 2 mm; (a) natural anterior view, and (b) posterior view. Labels 1–8 denote the entry and exit connections of the blood to the atrial cavities: (1) mitral valve; (2) tricuspid valve; (3) inferior vena cava; (4) superior vena cava; (5) left superior pulmonary vein; (6) right superior pulmonary vein; (7) left inferior pulmonary vein; (8) right inferior pulmonary vein.

a velocity of about 1 m/s. After recovery, the activation process is repeated only after a reinitiation in the SAN region.

In contrast, during AF parts of the myocardium are reactivated continuously by wavefronts previously set up in the tissue. This requires the incoming wavefront to arrive later than the local refractory period, a condition that is facilitated by any delay of the incoming wavefront or by any reduction of the local refractory period. The complex geometry of the atria (Figure 7.2) provides various possibilities for the formation of self-perpetuating processes around the orifices of the connecting vessels and those of the valves connecting atria and ventricles. The condition where a wavefront continuously circulates around such orifices, setting up a periodic, but highly abnormal, activation pattern at an unusually high rate, is called atrial flutter.

7.3.2   THE EQUIVALENT DOUBLE LAYER

During the activation and repolarization of the myocytes, electrical currents flow from the interior of an activated cell to those of its direct neighbors, passing through the intercalated disks. The return currents $I_m(t)$ that flow across the cellular membrane set up the potential field in the surrounding extracellular medium. These currents form the sources of observed electrograms (EGMs) and ECGs. By convention, a local outflow current $I_m(t)$ (of positive charge) is attributed a positive sign.

As described by Plonsey [11], a fundamental aspect of setting up a quantitative source description of the potentials at some distance from the active membrane is that these may (indeed) be described as *current* sources, and this in spite of the fact that their strength is derived from differences between the transmembrane *potentials* of neighboring cells.

The current source strength can be derived from the TMP by applying the bidomain theory [12, 8], which treats the myocardium as a macroscopic, homogenized medium. It uses a specification of an observation point in the intracellular and the extracellular medium by means of one and the same variable $\vec{r}$. In its monodomain approximation it is formulated as

$$I_m(\vec{r}, t) = \nabla \cdot \sigma_i(\vec{r}) \nabla V_m(\vec{r}, t) \approx \sigma_i \nabla^2 V_m(\vec{r}, t) , \tag{7.1}$$

where $I_m(\vec{r}, t)$ is an impressed current source volume density (unit: A m^{-3}), σ_i the electric conductivity of the intracellular domain, and $V_m(\vec{r}, t)$ the local TMP. ∇ denotes the vector operator for finding spatial derivatives (the gradient) and ∇^2 the Laplacian operator (computing the divergence of the gradient). The approximation shown assumes the electrical conductivity of the intracellular domain to be uniform.

A highly significant consequence of this expression is that it shows that, at all moments in time, the current sources are related to the second spatial derivative of the TMP $\nabla^2 V_m(\vec{r}, t)$, rather than to the TMP *per se*. This means that no external potential fields are generated by the activity at location $\vec{r}$ if $\nabla^2 V_m(\vec{r}, t)$ is zero, or more generally, if $\nabla \cdot \sigma_i(\vec{r}) \nabla V_m(\vec{r}, t)$ is zero. In case all myocytes are uniformly polarized the external field is zero. This is the condition usually assumed to prevail at the end of repolarization.

In the extracellular medium, taken here for the present moment to be of infinite size, with electrical conductivity σ_e, the potential $\Phi(\vec{y}, t)$ as generated by I_m as in (7.1) follows from solving the differential equation [13]

$$\sigma_e \nabla^2 \Phi(\vec{y}, t) = I_m(\vec{x}, t) , \tag{7.2}$$

which expresses the conservation of current. For reasons of clarity, $\vec{y}$ and $\vec{x}$ are used to denote the position of the field point and the source location, respectively. In the bidomain approach, for any point inside the myocardial wall we have: $\vec{y} = \vec{x} = \vec{r}$. For a field point outside the myocardium, with no so-called primary sources present, we have

$$\sigma_o \nabla^2 \Phi(\vec{y}, t) = 0 , \tag{7.3}$$

with σ_o the conductivity of the medium outside the heart.

A conceptually important first step in finding the potential field resulting from the current source is the computation of the so-called infinite medium potential. This is the potential field in the extracellular domain generated by the sources when placed inside a medium having an infinite extent and a homogeneous conductivity. When taking $\sigma_0 = \sigma_e$, the infinite medium potential solution to (7.2) and (7.3), after substitution of (7.1), is

$$\Phi_\infty(\vec{y}, t) = \frac{\sigma_i}{4\pi\sigma_e} \int_V \frac{\nabla^2 V_m(\vec{x}, t)}{r} \, dv , \tag{7.4}$$

with r the length of the vector $\vec{r}$ from any source location source $\vec{x}$ to field point $\vec{y}$. The volume integral in (7.4) is taken over all active sources within the entire atrial myocardium.

In spite of its simplicity, (7.4) is not very practical when computing the potential fields arising from a source distribution in three-dimensional space (the myocardium). The following, alternative expression was formulated by Geselowitz, [13, 14], who derived it by applying some classic results from field theory. It reads

$$\Phi_\infty(\vec{y}, t) = \frac{-\sigma_i}{4\pi\sigma_e} \int_S V_m(\vec{x}, t) \, d\omega \, , \tag{7.5}$$

with $d\omega = d\omega(\vec{y}, \vec{x})$ the solid angle subtended at field point $\vec{y}$ by an element around point $\vec{x}$ of the closed surface S bounding all atrial myocytes (endocardium and epicardium).

The expression in (7.5) shows that the integration over three-dimensional space involved in (7.4) may be replaced by integration over the bounding surface S, which has significant conceptual as well as computational advantages. The expression holds true for the monodomain approximation, provided the intracellular conductivity σ_i is taken to be uniform. No reliable experimental values of σ_i are currently available to permit a meaningful evaluation of the consequences of nonuniform tissue properties.

The nature of (7.5) permits the identification of an equivalent double layer (EDL) current source, located at S, with strength $\sigma_i V_m$. Its strength is expressed in units A/m=Am/m^2, which reveals its nature: a current dipole density per unit surface area.

The application of this source model requires the specification of the surface S as well as of the TMP at this surface.

7.4   THE VOLUME CONDUCTOR MODEL

The potential field generated inside the thorax and on its surface by myocardial sources is influenced by the electrical conductivities of the surrounding tissues. The latter are commonly modeled by nonintersecting compartments, each having a specific, homogeneous conductivity.

The effects of inhomogeneity can be computed by different methods. In our work, we use the boundary element method (BEM) [15, 16, 8]. This method computes the potential distribution throughout the body from

$$\Phi(\vec{r'}) = \frac{\sigma_s}{\sigma} \Phi_\infty(\vec{r'}) - \frac{1}{4\pi\sigma} \sum_\ell (\sigma_\ell^- - \sigma_\ell^+) \int_{S_\ell} \Phi_l(\vec{r}) \frac{\vec{R} \cdot d\vec{S}}{R^3} \, , \tag{7.6}$$

with

$\Phi_\infty(\vec{r'})$ the infinite medium potential, as, e.g., in (7.5),
σ_s the conductivity of the source region,
S_ℓ the interface bounding compartment l,
σ_ℓ^+ the conductivity just outside S_ℓ ($\ell = 1, N_S$) the number of interfaces,

σ_ℓ^- the conductivity just inside S_ℓ,
σ the conductivity at $\vec{r}'$, the observation point,
$\vec{r}$ a point at the interface S_ℓ,
$\vec{R}$ $= \vec{r}' - \vec{r}$; a vector of length R.

Equation (7.6) constitutes the essence of the BEM. It expresses the potential field in the "real world" medium as the potential inside a virtual infinite medium generated by the impressed primary sources (expressed by the first term on the right-hand side of (7.6)) and of virtual double layer sources at each of the interfaces bounding the subregions (second terms on the right-hand side). Note that the contributions of these secondary sources on any interface may be neglected if the difference between the conductivity values at both sides of the interface is small.

The major inhomogeneity involved is the torso boundary, where the conductivity drops to zero in the exterior region (air). Other major inhomogeneities that have been identified are those of the lungs and those resulting from the blood inside the ventricular and atrial cavities. The conductivity values of these compartments, relative to the mean conductivity value of the thorax, are 1/5 and 3, respectively. The geometry of these major compartments is shown in Figure 7.3.

7.4.1   POTENTIALS AT ARBITRARY FIELD POINTS

In practical applications to arbitrary field points, (7.6) can only be used if the potentials at the interfaces are known. These are computed by taking first the nodes at all interfaces as field points. The numerical implementation of this leads, for each of a set of N observation points placed on the interfaces, to a linear equation in the $n = 1, 2, \ldots, N$ unknown potentials. The total set of N equations in the N unknown values is solved by computer implementation of the appropriate methods of linear algebra.

The solution found is unique only up to a constant, corresponding to the nature of the physics of the problem: the gradient of the potential at the body surface is zero. This requires the specification of a point inside the medium, or on its boundary, to act as a potential reference. In this way a unique solution is specified, in the same manner as carried out while measuring bioelectrical potentials. The reference point may be chosen at will. As discussed in [17, 18], there is no natural, theoretical optimum location for the potential reference. The reference used in all examples shown in this chapter is the Wilson Central Terminal (WCT). This is an artificial reference, commonly used in electrocardiography, formed by averaging the potentials at both arms and the left foot.

The outcome of the entire procedure for determining all elements of an N-dimensional numerical vector may be formulated as a multiplication of the (numerical) vector of infinite medium potentials at arbitrary field points by a transfer matrix, say $\mathbf{A}$. The latter constitutes a numerically determined set of weighting coefficients, representing the effect of all volume conductor effects, i.e., the bounds of the volume conductor and its internal inhomogeneities. In this manner, both body surface potentials (ECGs) and anywhere inside the thorax (EGMs) can be computed.

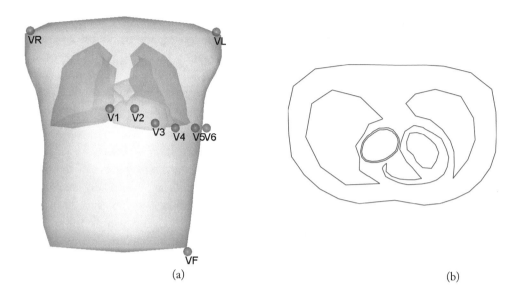

(a)   (b)

Figure 7.3: (a) Volume conductor model. Included: thorax boundary; atria, ventricles, and lungs. Geometry and locations of the nine electrodes of the standard 12-lead system, derived from MR images, are as indicated. (b) Transverse cross-section at the level of lead V1. Note: this level does not cross the left atrium in this particular subject, cf. Figure 7.2.

7.4.2   A GENERAL FORWARD FORMULATION

Both the source description and the volume conduction effects, as discussed above, may be expressed by means of matrices: the sources by a matrix $\mathbf{S}$, whose elements $s_{n,t}$ express the EDL source strength of node n of the atrial surface at time instant t, and a transfer matrix $\mathbf{A}$ with $a_{l,n}$ the potential generated at observation point l (electrode position) by a unit strength source element at node n:

$$\mathbf{\Phi} = \mathbf{AS} .$$ (7.7)

Recall that the EDL source strength on the surface S is proportional to the local TMP, cf. (7.5).

The boundary element method permits the computation of the potentials at any observation point inside the thorax. Observation points on the atrial surface result in EGMs; those on the torso surface result in ECGs.

7.5   MODELING ACTIVATION AND RECOVERY

As described in Section 7.3.1 the electrical forces driving the activation process are situated in, or near, the cell membrane. The EDL source description of atrial potentials in (7.5), as summarized in Section 7.3.2, requires the specification of the TMP at the surface S bounding the myocardium. These in turn stem from the sequence of depolarization and repolarization of the myocytes.

For modeling the activation sequence of the atria following the initiation at the SAN, a simplified approach based on the general principle of propagation following the Huygens principle has been shown to yield surprisingly realistic [19] and useful [20] results. However, for a description of the propagation and recovery during AF, the complexity involved demands a detailed description of the ion kinetic taking place at the cell membrane. This method is briefly summarized below.

7.5.1   PROPAGATION DERIVED FROM MEMBRANE KINETICS

The activation and recovery processes of the myocytes depend on the properties of the membrane channel kinetics as well as on the way in which the individual cells are electrically coupled to their direct neighbors. Morphological studies have yielded detailed information about cell coupling.

Mathematical modeling of the membrane kinetics has produced waveforms of TMPs of single cells that closely resemble those observed experimentally. For human atrial myocytes the most commonly used description of the ion kinetics is the model postulated by Courtemanche, Ramirez, and Nattel (the CRN model) [21]. It takes into account the instantaneous concentration of 12 different types of ionic channels, the variations in Ca^{2+}, Na^+, and K^+ intracellular concentrations, and the calcium dynamics in the sarcoplasmic reticulum. This is the kinetics model involved in the results described in this chapter.

The computation of $V_m(t)$ and $I_{ion}(t)$ for a single cell involves solving a system of 21 coupled nonlinear differential equations [21], for which standard numerical methods are available. Based on a specification of the spatial distribution of the myocytes and their coupling, the propagation may then be computed by solving the partial differential equation

$$C_m \frac{\partial V_m(\vec{r}, t)}{\partial t} = S_v^{-1} \, \nabla \cdot \sigma \, \nabla V_{m(\vec{r},t)} + I_{stim}(\vec{r}, t) - I_{ion}(\vec{r}, t) \,, \tag{7.8}$$

where $V_m(\vec{r}, t)$ denotes the TMP at location $\vec{r}$ at time t, C_m the membrane capacitance (unit: F m^{-2}), S_v the cell surface-to-volume ratio (unit: m^{-1}), and σ the electric conductivity of the medium (unit: S m^{-1}). The surface-to-volume ratio S_v represents the total area of cell membrane per unit of tissue volume.

The system is driven by the ionic current density $I_{ion}(\vec{r}, t)$ (unit: A m^{-2}) generated by the active processes near the membranes and are computed by solving, for each cell considered, the CRN kinetics equations. Any external stimulus activating the tissues is represented by the impressed current density $I_{stim}(\vec{r}, t)$ (unit: A m^{-2}).

Equation (7.8) constitutes a so-called reaction diffusion equation where the term $S_v^{-1} \, \nabla \cdot \sigma \, \nabla V_m(\vec{r}, t)$, i.e., the diffusion term, represents the electrical current flowing into the cell as a result of the intercellular coupling between neighboring cells and their potential differences. Equation (7.8) is expressed in the so-called monodomain approximation of the bidomain formulation governing the propagation of the cardiac impulse in the myocardium [12, 8].

7.5.2   UNITS REPRESENTING THE MYOCARDIAL CELLS

Equation (7.8) is expressed in continuous variables of space and time. The numerical handling of this expression demands the introduction of a spatial discretization. This is carried out by introducing so-called units, each representing the activity of a conglomerate of myocytes. The large number of myocytes involved in any tissue model, for the atria alone as many as $\approx 2 \times 10^8$, the different types of ions, ion channels, and their numerous gating parameters, prohibit the formulation of a fully comprehensive number of units. The computer models of electrical propagation in the atria [22, 23, 24, 25, 26, 27, 28] differ in their representation of electrophysiological and atrial anatomical details. Typically, rather than dealing with the $\approx 2 \times 10^8$ myocytes of the atria, a far smaller number of coupled units have been used in this type of research, as many as 4,000,000 in some of these.

Propagation of activation relies on the interaction of neighboring myocytes and/or specialized cells like those in the SAN region and those of the AV node. Different types of mathematical methods have been worked out for simulating normal propagation processes as well as for studying the conditions that give rise to fibrillatory, chaotic-like behavior of the entire system.

7.6   MODEL PARAMETERS AS USED IN THE EXAMPLES

The specifications of the model and its parameters used in the examples presented in the subsequent sections are as follows. The geometry involved was from magnetic resonance (MR) images of healthy subjects. Details of the collection and basic processing of these data are included in the cited references to earlier publications. In addition to the introductory treatment in the previous sections, this section specifies some of the details of the models used in the simulation of the various signals shown in the final section.

7.6.1   ATRIAL MODEL

The atrial model was derived from data first published in the form of a mono layer [27]. Its geometry was specified by about 100,000 nodes. Subsequently, based on this mono layer, a thick-wall variant was constructed, having a thickness in the range of 1.2–1.5 mm. The boundary was represented by a triangulated surface, specified by 1,297 nodes. Within the resulting, closed hull bounding the atrial myocardium, some 800,000 units were placed in a locally cubical array, with spacing of 0.3 mm [20]. Details of the geometry are indicated in Figure 7.2.

Simulated Normal Propagation

Propagated activation and recovery was simulated by means of the method described in Section 7.5.1. Standard settings for the parameters of CRN membrane were taken from the literature [21]. The settings of the parameters in (7.8) were: $C_m = 1 \ \mu F/cm^2$, $S_v = 0.24 \ \mu m^{-1}$, and $\sigma = 0.56$ S/m (uniform, isotropic) in the bulk of the tissue. Equation (7.8) was solved numerically on a cubical grid in the interior of the atrial wall, using a finite difference approach [29], with time steps ranging from 12.5–50, adapted to the dynamical state of the ion kinetics [30]. At the surface bounding the myocytes, the normal spatial derivative of the TMP was used as the boundary condition. Following the application of a stimulus applied in the region of the SAN, a single wavefront propagated over the atrial surface at a velocity of 0.7 m/s. The activation sequence agreed qualitatively with the data reported in the literature for normal atria.

Substrate for AF

When using a uniform setting for all parameters of the kinetics model, it was found that AF-like activity could not be introduced. Following any rapid pacing, which introduced and maintained a chaotic type of atrial activation, the system invariably returned to a fully polarized when the stimulus ended. In order to create a substrate for AF some of the parameters of the CRN kinetics were modified. These were aimed at creating patchy heterogeneities (at a spatial scale of about 2 cm), affecting the intrinsic value of the local action potential duration APD$_{90}$ [31]. To this end, the channel conductances associated with the ionic currents I$_{to}$, I$_{CaL}$, I$_{Kur}$, and I$_{Kr}$ were scaled with respect to those used for normal propagation. Two different sets of values were used. In set A, the above parameters were scaled with respect to the normal settings by factors 3, 0.5 , 1, and 3, respectively. These values were assigned to the major part of the tissue. For the second set, B, these scaling factors were 0.2, 0.7, 0.1, and 1.5, respectively. The values of set A were assigned to the major part of the tissue, those of set B to the patches. With these settings, when stimulated from the SAN area at a rate of 120 bpm, the APD$_{90}$ values observed over the entire surface showed a range of 150–230 ms (mean: 195±15 ms). For these inhomogeneous parameter settings, sustained AF was induced by rapid pacing (cycle length: 150 ms), with the stimulus applied to the appendage of the left atrium.

7.6.2   THORAX MODEL

The geometry of a healthy subject was documented by MR imaging methods [32]. The MR images were used to extract the geometry of the heart, lungs, and thorax boundary, as well as of the ventricular cavities. To this ventricular geometry, tested extensively in previous studies, the geometry of the atrial model (Figure 7.2) was fitted by means of translation, rotation, and scaling. The atrial cavities were connected to the corresponding ventricular ones. These cavities constitute the major volume conduction inhomogeneities inside the thorax.

The electrical conductivity was set at 0.2 S/m in the torso and in the myocardium, at 0.6 S/m in the blood filled cavities, and at 0.04 S/m in the lungs. The electrical conductivity was assumed to be homogeneous and isotropic inside each of the compartments [33]. Finally, the potentials at all conductivity interfaces were computed by means of dedicated BEM software, developed during previous studies [32]. This is the geometry shown in Figure 7.3.

7.7   EXAMPLES

Different validation studies on the model based approaches to the genesis of atrial signals have been carried out in our group. For the genesis of the normal P wave these can be found in [20], for AF, see, e.g., [27, 6]. Studies in which the model has been applied include [31, 6, 2, 34].

Below, we present some applications of the model to describing the genesis of AF signals. The nature of the model permits us to carry out this analysis at the full spatiotemporal level of the problem involved. The signals illustrated are taken from one and the same episode of stable AF simulated over a period of 30 s.

7.7.1   POTENTIALS ON THE ATRIAL SURFACE

Potential Fields

In this chapter different types of scalar fields have been implied. In Section 7.3.2 the TMP over the atrial surface bounding the myocardium, S, was introduced, acting as the EDL source strength. An example of this distribution at an arbitrary point in time during a period of simulated AF is shown in Figure 7.4(a). At this particular moment, several different regions can be distinguished, relating to different stages of activation and recovery. Narrow bands of TMPs of about -35 mV can be observed, e.g., the one on the right atrium moving toward the right atrial appendix. Such narrow bands correspond to depolarization wavefronts moving toward the bordering, polarized zone. The magnitude of the jump in the TMP while moving across the wavefront is about 80 mV. The more spread-out band of TMPs around -35 mV corresponds to repolarization, migrating toward the depolarized zone with near zero instantaneous TMPs.

The potential field over S (encompassing epicardium and endocardium) has a different nature: rather than being a potential difference across the membranes of the local myocytes, it is the potential in the extracellular medium, referred to the external reference used, i.e., the WCT. The potential field generated by the source strength depicted in Figure 7.4(a) is shown in panel (b) of the same figure. By comparing both panels, the major difference observed is a reversal of the polarity of the potential gradient across the wavefronts. On the epicardium, ahead of the wavefront the potentials are about 10 mV higher than behind it. On the endocardium the jump is somewhat smaller. The spatial mean of the TMP over the entire surface S varies with time (mean: -54 mV, range: -61 to -45 mV). For this time instant, as well as for all other, the spatial mean of the simulated external potential on S is zero. This relates to the fact that the spatial mean of the EDL source (7.5) constitutes a closed uniform double layer which has the property of producing no external field [35, 9].

The reversal of polarity corresponds to the fact that the external potentials are generated by the return of the activating currents that—in the intracellular domain—flow toward the region still at rest (polarized).

The differences in the potential jumps across the wavefront, -80 mV versus 10 mV, can be explained on the basis of (7.5) and (7.6). When the wavefront in the tissue passes the external observation point on S, the solid angle jumps by 2π. Inserted in (7.5) this would "predict" a jump of $40\sigma_i/\sigma_e$ mV in the external field. The order of magnitude of the 10 mV jump in the external medium may be checked by taking realistic estimates of the various conductivities, and taking into account the volume conduction effects due to the three-fold higher conductivity of the blood inside the cavities, computed by using (7.6). These inhomogeneities also explain the smaller values of the

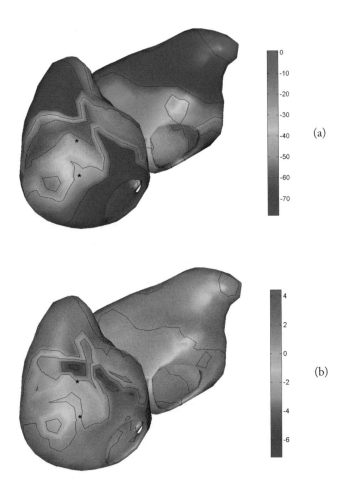

Figure 7.4: (a) Instantaneous distribution of the local TMPs (the EDL source strength over the atrial surface) during AF. Anterior view of the atrial model; Isopotential line increments: 10 mV; the narrow band of potentials of about −35 mV corresponds to depolarization wavefronts moving toward the bordering polarized with TMPs of −70 mV, the wider band to regions in which repolarization migrates toward the bordering depolarized zone of near zero TMPs. (b) Potential distribution on the atrial surface S generated by the source distribution shown in (a); increments between isopotential lines: 2 mV. The lower of the two dots drawn on the right atrium mark the position of node $n1$, the node closest to the position of electrode V1 on the thorax, the other to a nearby node, $n2$, at a distance of 1.8 cm. Electrograms at these locations are shown in Figure 7.6.

jumps on the endocardium relative to those on the epicardium. In "real life" situations the actual scaling factors involved may also differ due to local differences in the thickness of the atrial wall.

Signals

Although the patterns shown in Figure 7.4 are similar, they are by no means identical. In fact, the spatial correlation coefficients between these fields, computed over a period of 2 s, showed values ranging between −0.79 and −0.52 (mean: −0.70). This corresponds to their different nature: the TMP values shown depend, exclusively, on the *local* source strength, but those of the potential field are generated by all sources distributed over S.

These differences are immediately evident when comparing the source strength and the external potential as a function of time. The two upper traces of Figure 7.5 serve for such a comparison. The upper trace depicts the TMP at an epicardial node (node $n1$) indicated in Figure 7.4. The middle trace shows the EGM (simulated) "recorded" directly at the same node. The potential reference is the WCT. In this EGM, the passing of a wavefront below the electrode is dominated by the fast downward slope. This passage is a local phenomenon that is surrounded by large spatial gradients. The magnitude of the fast downward signal component is about 10 mV. The starting level of this component, the amplitude of the "R wave," and consequently also the value of the "S wave," depend on the more diffuse contributions of all source elements over S.

Following the passage of the wavefront, some expression of local repolarization can be seen in the EGM. Since the local, spatial gradients during this stage are much smaller, the signal values are also much smaller. During AF, there is an ongoing activity of distal sources, which prevents an accurate, direct interpretation of local repolarization on the basis of just a single EGM.

The signal shown in Figure 7.5 may be classed an EGM taken under ideal conditions: the electrode was in direct contact with the epicardium. When the distance between sensing electrode and active tissue increases, the expression of the passing wavefront gets more diffuse, the amplitudes of the signals are reduced, and their interpretation more difficult.

In clinical practice, the EGM shown in Figure 7.5 is referred to as a "unipolar" lead. This is in fact a confusing, and basically incorrect, term: the recording of any potential difference always requires at least two poles, or terminals. In addition to unipolar leads, so-called "bipolar" leads are recorded, recordings of the potential difference between two closely spaced electrodes. This electrode configuration is more sensitive to the contributions of local sources than to those of distal sources, all the more so the smaller the interelectrode distance. As such it may be viewed as a spatial filter. Unfortunately, this filter does not just reduce the signal components generated by distal sources: in applications to the atria, e.g., those of the contaminating ventricular sources. The filter also affects the contributions of the more local sources, both in terms of their amplitude (a reduction) and their waveform.

The model enables a comprehensive analysis of such effects. Here, a single example is presented. In Figure 7.6, the top panel shows the paired TMP and EGM signals at node $n1$, the same as the ones shown by the upper and middle traces of Figure 7.5. Below, a similar pair is shown, recorded at another node $n2$, located on the epicardium of the right atrium (Figure 7.4) at a distance

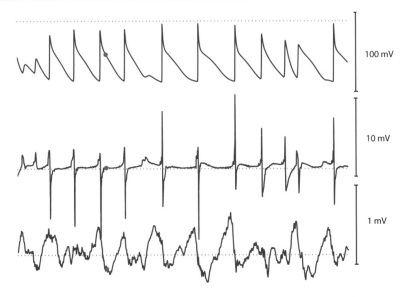

Figure 7.5: AF signals simulated over a 2-s period. Upper trace: TMP at the epicardial node (node $n1$) of the atrial model closest to electrode V1 on the thorax (Figure 7.4). Middle trace: Electrogram at node $n1$. Lower trace: corresponding ECG signal of lead V1. For all signals, the dotted horizontal lines denote zero potential. The heavy dots mark the time instant for which locations the EDL source distribution is shown in the left panel of Figure 7.4.

of 1.8 cm from node $n1$. The lower panel of Figure 7.6 depicts the corresponding bipolar lead, the signal that results by subtracting the unipolar EGM at node $n2$ (middle panel) from the one at node $n1$ (upper panel). With the unipolar signals at $n1$ and $n2$ available, the interpretation of the bipolar signals is straightforward. Because of the wiring of the bipolar lead, a fast negative deflection in the bipolar signal corresponds to the passage of a wavefront at node $n1$, and a fast positive deflection to the passage of a wavefront at $n2$. If a single wavefront would pass both electrodes simultaneously, the line connecting the electrodes being parallel to the wavefront, the deflection in the bipolar lead will be close to zero. In this situation, the timing of local activation can not be extracted from the bipolar lead signal.

7.7.2   POTENTIALS ON THE THORAX SURFACE

Similar to what is shown in Figure 7.4(b) for the atrial surface, the potentials on the thorax, as generated by atrial sources during AF, may be visualized in the form of maps, the body surface potential maps (BSPMs). In Figure 7.7(a) the BSPM is shown, generated at the time instant for which the (AF) source distribution is depicted in Figure 7.4(a). By comparing Figure 7.7(a) with Figures 7.4(a) and (b), a major influence of the volume conductor effects can be observed. The image

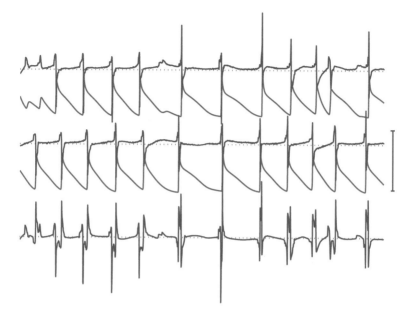

Figure 7.6: AF signals simulated over a 2-s period. Upper panel: superimposition of the TMP and the electrogram at epicardial node $n1$ (Figure 7.4). This pair replicates the upper two traces of Figure 7.5. Middle panel: Superimposition of the TMP and the EGM at an epicardial node $n2$, a node at about 1.8 cm from $n1$. Lower panel: bipolar EGM resulting from subtracting the monopolar EGM at node $n2$ from that at node $n1$. Vertical scale as in Figure 7.5.

of the source distribution at the atrial surface S appears in a highly diffuse, convoluted manner on the body surface.

The location of the precordial electrodes of the standard 12-lead system can be seen to give nearly the same instantaneous potential, and as such can be seen to be suboptimal for spatially sampling the potential field during AF.

7.7.3 ECG SIGNALS

The more common way of looking at body surface potentials is by means of the standard 12-lead signals. Lead V1 of the simulated AF signal during the 2-s period is depicted by the lower trace of Figure 7.5. The similarity between this signal and the TMP (upper trace) of the nearest source element $n1$ of S is clearly "not very high." The distance between the locations of electrode V1 and $n1$ was about 7 cm. Since all other (1296) source elements were located at distances of no more than 17 cm away, their simultaneously ongoing activity mingled with that of $n1$.

The corresponding, complete set of standard 12-lead signals is shown in Figure 7.8. In this figure, the extremity leads shown are VR, VL, and VF rather than the limb leads aVR, aVL, and aVF.

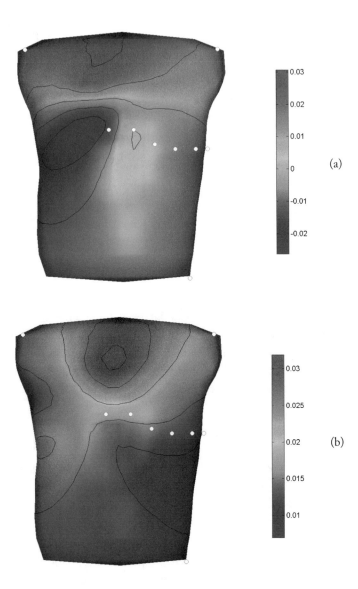

Figure 7.7: (a): Body surface potential map generated by the EDL at the time instant for which the source distribution is shown in Figure 7.4(a). Isopotential lines drawn at increments of 10 μV. (b) Map of the root mean-square values of simulated AF, computed over a 2-s period. Isofunction lines drawn at increments of 5 μV. Potential reference: Wilson central terminal.

This unaugmented scaling puts these signals on equal footing with the precordial leads V1, . . .,V6. The augmented signals are scaled up by a factor of 1.5 as a result of the wiring of the electrodes involved, a property that was very useful in the early days of electrocardiography, when electronic amplification was not available.

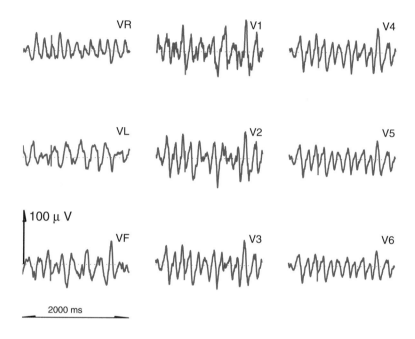

Figure 7.8: Standard 12-lead ECG signals of simulated AF signals. Extremity leads VR, VL, and VF unaugmented (see text). The small vertical bar shown in each signal marks the timing of the source strength shown in Figure 7.4(a). Time interval shown: 2 s.

Because of their chaotic background, AF signals cannot be expressed by their amplitudes. Instead, their magnitudes may be specified by the root mean square (RMS) signal values computed over some time interval. The model based approach permits the computation of the RMS values of signals simulated at any (electrode) location on the thorax. A map of these values computed from AF signals over a 2-s interval is presented in Figure 7.7(b). Note that the RMS values at the location of the precordial leads V1, . . .,V6 are far smaller than those in the upper sternal region.

In another application of the model, the possibility of significant correlations between precordial signals, as suggested by inspection of BSPMs like the one in Figure 7.7(a), was confirmed by computing the matrix of the (estimated) linear correlation coefficients $\hat{\rho}$ of these signals. The $\hat{\rho}$ values of neighboring leads showed values ranging from 0.75 ($\hat{\rho}(V1,V2)$) to 0.99 ($\hat{\rho}(V5,V6)$). Such highly correlated signals can not be expected to yield independent information on AF, suggesting that the extraction of information on AF from the basis of the electrode positions of the standard

12-lead system is suboptimal. Ideally, much denser spatial sampling of the region between the level of V1 and the upper sternal region should be used. If just the nine electrodes of the standard 12-lead system are available, an adapted electrode configuration can be applied [36].

Bibliography

[1] W. Einthoven and K. de Lint, "Ueber das normale menschliche Elektrokardiogram und die capillar-elektrometrische Untersuchung einiger Herzkranken," *Pflügers Arch. ges Physiol.*, vol. 80, pp. 139–160, 1990.

[2] M. Lemay, J.-M. Vesin, A. van Oosterom, V. Jacquemet, and L. Kappenberger, "Cancellation of ventricular activity in the ECG: evaluation of novel and existing methods," *IEEE Trans. Biomed. Eng.*, vol. 54, pp. 542–546, 2007. DOI: 10.1109/TBME.2006.888835

[3] V. Barbaro, P. Bartolini, G. Calcagnini, and F. Censi, "Extraction of physiological and clinical information from intra-atrial electrograms during atrial fibrillation: review of methods," *Ann. Ist. Super Sanita*, vol. 37, pp. 319–324, 2001.

[4] K. T. Konings, C. J. Kirchhof, J. R. Smeets, H. J. Wellens, O. C. Penn, and M. A. Allessie, "High-density mapping of electrically induced atrial fibrillation in humans," *Circulation*, vol. 89, pp. 1665–1680, 1994.

[5] S. Shpun, L. Gepstein, G. Hayam, and S. A. Ben-Haim, "Guidance of radiofrequency endocardial ablation with real-time three-dimensional magnetic navigation system," *Circulation*, vol. 96, pp. 2016–2021, 1997.

[6] V. Jacquemet, A. van Oosterom, J.-M. Vesin, and L. Kappenberger, "Analysis of electrocardiograms during atrial fibrillation. A biophysical model approach," *IEEE Eng. Med. Biol. Mag.*, vol. 25, pp. 79–88, 2006. DOI: 10.1109/EMB-M.2006.250511

[7] R. M. Gulrajani, F. A. Roberge, and G. E. Mailloux, "The forward problem of electrocardiography," in *Comprehensive Electrocardiology* (P. Macfarlane and T. Lawrie, eds.), pp. 197–236, Oxford: Pergamon Press, 1989.

[8] R. M. Gulrajani, *Bioelectricity and Biomagnetism*. New York: John Wiley & Sons, 1998.

[9] R. Plonsey and R. Barr, *Bioelectricity: A Quantitative Approach*. New York: Springer, 3rd ed., 2007.

[10] H. Snellen, *Selected Papers on Electrocardiography of Willem Einthoven*. Leiden: Leiden University, 1997.

[11] R. Plonsey, "An extension of the solid angle potential formulation for an active cell," *Biophys. J.*, vol. 5, pp. 663–667, 1965.

[12] C. S. Henriquez and A. A. Papazoglou, "Using computer models to understand the roles of tissue structure and membrane dynamics in arrhythmogenesis," *Proc. IEEE*, vol. 84, pp. 334–354, 1996. DOI: 10.1109/5.486738

[13] D. B. Geselowitz, "On the theory of the electrocardiogram," *Proc. IEEE*, vol. 77, pp. 857–876, 1989. DOI: 10.1109/5.29327

[14] D. B. Geselowitz, "Description of cardiac sources in anisotropic cardiac muscle. Application of bidomain model," *J. Electrocardiol.*, vol. 25 Suppl, pp. 65–67, 1992. DOI: 10.1016/0022-0736(92)90063-6

[15] R. C. Barr, M. Ramsey, and M. S. Spach, "Relating epicardial to body surface potential distributions by means of transfer coefficients based on geometry measurements," *IEEE Trans. Biomed. Eng.*, vol. 24, pp. 1–11, 1977. DOI: 10.1109/TBME.1977.326201

[16] A. van Oosterom, "Electrocardiography," in *The Biophysics of Heart and Circulation* (J. Strackee and N. Westerhof, eds.), pp. 249–256, Bristol: Inst of Physics Publ, 1993.

[17] H. C. Burger, "The zero of potential: A persistent error," *Am. Heart J.*, vol. 49, pp. 581–586, 1955. DOI: 10.1016/0002-8703(55)90076-4

[18] D. B. Geselowitz, "The zero of potential," *IEEE Eng. Med. Biol. Mag.*, vol. 17, pp. 128–136, 1998. DOI: 10.1109/51.646230

[19] P. M. van Dam and A. van Oosterom, "Atrial excitation assuming uniform propagation," *J. Cardiovasc. Electrophysiol.*, vol. 14, pp. S166–S171, 2003. DOI: 10.1046/j.1540.8167.90307.x

[20] A. van Oosterom and V. Jacquemet, "Genesis of the P wave: Atrial signals as generated by the equivalent double layer source model," *Europace*, vol. 7 Suppl 2, pp. 21–29, 2005. DOI: 10.1016/j.eupc.2005.05.001

[21] M. Courtemanche, R. J. Ramirez, and S. Nattel, "Ionic mechanisms underlying human atrial action potential properties: Insights from a mathematical model," *Am. J. Physiol.*, vol. 275, pp. H301–H321, 1998.

[22] D. Harrild and C. Henriquez, "A computer model of normal conduction in the human atria," *Circ. Res.*, vol. 87, no. 7, pp. E25–E36, 2000.

[23] C. J. Kafer, "Internodal pathways in the human atria: A model study," *Comput. Biomed. Res.*, vol. 24, pp. 549–563, 1991. DOI: 10.1016/0010-4809(91)90039-Y

[24] E. Macchi, "Digital-computer simulation of the atrial electrical excitation cycle in man," *Adv. Cardiol.*, vol. 10, pp. 102–110, 1974.

[25] G. K. Moe, W. C. Rheinboldt, and J. A. Abildskov, "A computer model of atrial fibrillation," *Am. Heart J.*, vol. 67, pp. 200–220, 1964. DOI: 10.1016/0002-8703(64)90371-0

[26] E. J. Vigmond, R. Ruckdeschel, and N. Trayanova, "Reentry in a morphologically realistic atrial model," *J. Cardiovasc. Electrophysiol.*, vol. 12, pp. 1046–1054, 2001. DOI: 10.1046/j.1540-8167.2001.01046.x

[27] N. Virag, V. Jacquemet, C. S. Henriquez, S. Zozor, O. Blanc, J.-M. Vesin, E. Pruvot, and L. Kappenberger, "Study of atrial arrhythmias in a computer model based on magnetic resonance images of human atria," *Chaos*, vol. 12, pp. 754–763, 2002. DOI: 10.1063/1.1483935

[28] C. W. Zemlin, H. Herzel, S. Y. Ho, and A. Panfilov, "A realistic and efficient model of excitation propagation in the human atria," in *Computer Simulation and Experimental Assessment of Cardiac Electrophysiology* (N. Virag, O. Blanc, and L. Kappenberger, eds.), pp. 29–34, Armonk, New York: Futura Publishing, 2001.

[29] F. Fenton and A. Karma, "Vortex dynamics in three-dimensional continuous myocardium with fiber rotation: Filament instability and fibrillation," *Chaos*, vol. 8, pp. 20–47, 1998. DOI: 10.1063/1.166311

[30] Z. Qu and A. Garfinkel, "An advanced algorithm for solving partial differential equation in cardiac conduction," *IEEE Trans. Biomed. Eng.*, vol. 46, pp. 1166–1168, 1999. DOI: 10.1109/10.784149

[31] V. Jacquemet, N. Virag, and L. Kappenberger, "Wavelength and vulnerability to atrial fibrillation: Insights from a computer model of human atria," *Europace*, vol. 7 Suppl 2, pp. 83–92, 2005. DOI: 10.1016/j.eupc.2005.03.017

[32] G. Huiskamp and A. van Oosterom, "The depolarization sequence of the human heart surface computed from measured body surface potentials," *IEEE Trans. Biomed. Eng.*, vol. 35, pp. 1047–1058, 1988. DOI: 10.1109/10.8689

[33] A. van Oosterom, "Genesis of the T wave as based on an equivalent surface source model," *J. Electrocardiol.*, vol. 34 Suppl, pp. 217–227, 2001. DOI: 10.1054/jelc.2001.28896

[34] A. van Oosterom, Z. Ihara, V. Jacquemet, and R. Hoekema, "Vectorcardiographic lead systems for the characterization of atrial fibrillation," *J. Electrocardiol.*, vol. 40, pp. 343.e1–343.11, 2007.

[35] F. N. Wilson, A. G. Macleod, and P. S. Barker, "The distribution of action currents produced by the heart muscle and other excitable tissues immersed in conducting media," *J. Gen. Physiol*, vol. 16, pp. 423–456, 1933. DOI: 10.1085/jgp.16.3.423

[36] Z. Ihara, A. van Oosterom, V. Jacquemet, and R. Hoekema, "Adaptation of the standard 12-lead electrocardiogram system dedicated to the analysis of atrial fibrillation," *J. Electrocardiol.*, vol. 40, pp. 68.e1–68.e8, 2007. DOI: 10.1016/j.jelectrocard.2006.04.006

CHAPTER   8

Algorithms for Atrial Tachyarrhythmia Detection for Long-Term Monitoring with Implantable Devices

Rahul Mehra, Jeff Gillberg, Paul Ziegler, and Shantanu Sarkar

8.1   INTRODUCTION

Atrial tachyarrhythmias are a significant component of the global burden of cardiovascular disease and its incidence is increasing rapidly due to the aging of the population. Atrial tachyarrhythmias result in poor quality of life for patients as they cause symptoms of tiredness, palpitations, and dizziness although, in many patients, episodes of atrial tachycardia/atrial fibrillation (AT/AF) can be completely asymptomatic. Whether it is symptomatic or asymptomatic, AT/AF also increase the risk of stroke in the presence of other clinical factors such as age above 75, diabetes, hypertension, heart failure, or prior stroke. It is estimated that one out of every six strokes occurs in patients with AT/AF [1]. Atrial tachyarrhythmias lead to more hospital admissions than any other arrhythmia [2], and the mortality rate in patients with AF is twice that of patients in normal sinus rhythm (NSR) [3]. Clearly, long-term prevention of AT/AF requires significant changes in lifestyle habits that impact risk factors such as hypertension and heart disease. However, in the short-term, management of these arrhythmias with pharmacologic agents, ablation, and implantable devices will continue to play a critical role in improving symptoms and reducing the risk of hospitalization and stroke.

In order to optimize management, the atrial tachyarrhythmia needs to be detected, monitored, and its characteristics quantify as per their frequency, duration, and effect on the ventricular rate. Firstly, the physician needs to evaluate with adequate monitoring if the patient's symptoms are caused by AT/AF. If they are, the patient is managed with one of two strategies: rhythm or rate control [4]. The goal of rhythm control is to reduce the total duration (burden) of AT/AF that a patient has over a given period of time, whereas rate control requires maintenance of physiologically appropriate ventricular rates during the AT/AF. Rapid rates during AT/AF can result in severe symptoms, while maintaining the ventricular rate within certain limits during AT/AF is associated with improvement of these symptoms. Both of these strategies have been investigated in clinical trials with similar outcomes [5]. Secondly, if the patient has symptomatic or asymptomatic AT/AF episodes and clinical risk factors for stroke, anticoagulation therapy needs to be initiated to reduce this risk.

8.1.1   EXTERNAL VERSUS IMPLANTABLE DEVICES

Measuring the characteristics of the atrial tachyarrhythmia is a challenge as the symptoms do not always correlate with the presence of an arrhythmia and many episodes can be completely asymptomatic [6]. Monitoring of AT/AF has traditionally relied on intermittent use of external devices. These snapshot monitoring techniques include routine ECG examination at clinic visit, 24–48 h Holter monitoring, or several days of monitoring triggered by patient symptoms, i.e., event monitoring. Newer systems, such as mobile cardiac outpatient telemetry, are capable of monitoring patients continuously for up to 30 days [7] and have been shown to increase the yield for arrhythmia detection compared with a single 24-h Holter monitor [8]. However, the ability of these external monitors to quantify the arrhythmia is significantly compromised. In a recent study, we compared the ability to identify patients with AT/AF via symptoms, intermittent external monitoring, and continuous monitoring with implantable devices [9]. Symptom-based and intermittent external monitoring methods were shown to have significantly lower sensitivity (31–71%) and negative predictive value (21–39%) for identification of patients with AT/AF (Figure 8.1) and significantly underestimated AT/AF burden compared to continuous monitoring. Additionally, these external devices interfere with showering and other daily activities, and the patch electrodes can cause skin irritation over such prolonged usage. Consequently, patient compliance with these systems can be quite low [7, 10].

Since external monitoring can only be performed intermittently over time, the likelihood of missing paroxysmal episodes of asymptomatic AT/AF is quite high. The incidence of asymptomatic AT/AF has been reported to be 20–50% [11, 12, 13, 14, 15, 16], and its presence poses a high unperceived stroke risk as was shown in the RACE [17] and AFFIRM [5] trials. Results from these studies showed that there was a high stroke rate in patients who were thought to be well rhythm controlled based on symptoms and infrequent monitoring. Therefore, continuous monitoring of rhythm control is critically important for managing a patient's anticoagulation regimen [9]. Two key clinical questions that arise are "when does a patient need to initiate anticoagulation therapy?" and "when is it safe to discontinue anticoagulation therapy?" Current guidelines recommend anticoagulation for patients with risk factors and AT/AF, regardless of the amount of AT/AF [4]. Therefore, it is important to identify patients who experience even relatively brief episodes of AT/AF.

8.1.2   IMPLANTABLE DEVICES

One strategy to address this need for continuous monitoring is to develop small implantable devices (similar to Reveal Plus, Medtronic Inc., displayed in Figure 8.2(a)) that would be capable of moni-

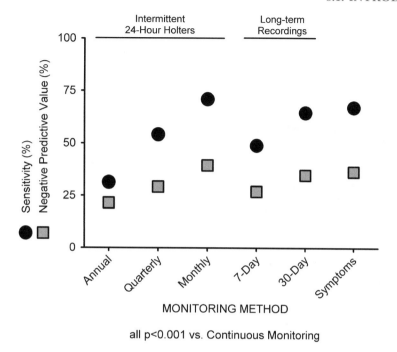

Figure 8.1: Sensitivity and negative predictive value for identification of patients with any AT/AF episodes detected by various intermittent external monitoring methods or symptoms.

toring atrial arrhythmias continuously over the lifetime of the device with both high sensitivity and specificity for AT/AF detection [18, 19]. Although these devices require an invasive implant procedure, patient compliance is not an issue as these devices rarely interfere with normal daily activities. The Reveal device is presently being implanted primarily for diagnosis of syncope. Patients are able to record the occurrence of symptoms within the memory of the device with an external activator; the resulting data can be retrieved later by the clinician. This device, about the size of a cigarette lighter, does not have intracardiac leads as the bipolar recording electrodes are located on the surface of the device itself. These devices are typically implanted in the subcutaneous space on the thorax of the patient. The characteristics of the sensed signals, as well as the detection algorithms of these devices, are significantly different from those with intracardiac leads. With a subcutaneously implanted Reveal, P waves are of very small amplitude and do not provide good signal to noise discrimination. Therefore, the detection algorithms are based on the sensing of R waves, a challenge for detection of atrial tachyarrhythmias; such algorithms will be discussed later in Section 8.4. Once the arrhythmia has been well classified, pharmacologic and ablation therapies are still the primary modalities for its management. The reported efficacy of clinical procedures such as pulmonary vein ablation for the treatment of AT/AF can vary greatly depending on how the arrhythmia is monitored [10].

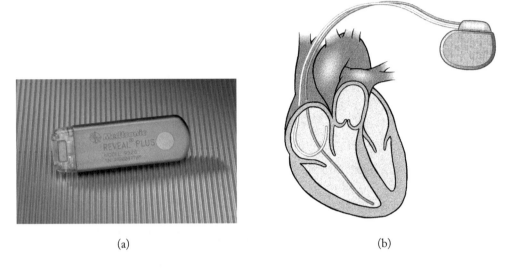

(a)　　　　　　　　　　　　　　　　　　　　　(b)

Figure 8.2: (a) Reveal device (Medtronic Inc.) with two subcutaneous electrodes. One electrode is circular (on the right of "Plus") and the other one rectangular on the header block, being about 2.5 cm apart. (b) A dual chamber implantable pacemaker with one lead in the right atrium and the other lead in the right ventricle.

Such results highlight the clinical need for continuous monitoring that does not rely upon patient compliance.

Presently, implanted devices play a key role in chronic monitoring for management of AT/AF primarily in patients who get implanted with a pacemaker or an implantable cardioverter defibrillator (ICD) for the management of their bradyarrhythmias, ventricular tachyarrhythmias, or heart failure, and in whom atrial tachyarrhythmias are a co-morbidity. About 30–50% of patients with devices for these indications have AT/AF and the implantable devices need to detect this arrhythmia for its treatment or prevention and to ensure proper operation of the device. Therefore, apart from the reasons stated above for chronic AT/AF monitoring, other clinical applications of detection/sensing algorithms include:

1. ensuring appropriate timing of pacing stimuli in the atrium and ventricle with sensing and modeswitching algorithms;

2. delivering appropriate atrial antitachyarrhythmia therapy (pacing or defibrillation) with algorithms that detect atrial tachyarrhythmias in devices that conduct rhythm control; and

3. delivering appropriate ventricular antitachyarrhythmia therapy in ICDs with algorithms that discriminate ventricular from atrial tachyarrhythmias.

These clinical applications and the function of sensing and detection algorithms in various device families are shown in Tables 8.1 and 8.2, respectively. Most such devices implanted today are dual chamber devices, i.e., have intracardiac leads for detection of atrial and ventricular signals (Figure 8.2(b)). These endocardial electrodes make good contact with cardiac tissue and can electrically stimulate it to initiate a contraction. They also record intracardiac electrograms (EGMs) from the atrium and the ventricle to sense the P and R waves. The sensed waves are then analyzed by the detection algorithm. Algorithms for AT/AF detection are designed based on the application for which they will be used. For example, in devices that provide atrial therapies, AT/AF detection should be highly sensitive to longer duration episodes and specific so that inappropriate therapies are not delivered. In a pacemaker, AT/AF detection needs to be very sensitive to short duration episodes because the operation (mode) of a pacemaker must be changed from synchronizing the ventricular response during normal atrial activity to operating asynchronously during AT/AF.

To discuss all these aspects, this chapter has been divided into four sections: (a) sensing and detection for implantable pacemakers, ICDs, and cardiac resynchronization therapy (CRT) devices; (b) detection of atrial tachyarrhythmias by pacemakers, ICDs, and CRT devices; (c) atrial tachyarrhythmia detection with subcutaneous monitoring devices; and (d) monitoring and clinical diagnostics for atrial tachyarrhythmias with implantable devices.

8.2 SENSING AND DETECTION FOR IMPLANTABLE PACEMAKERS, ICDS, AND CRT DEVICES

One of the primary functions of a pacemaker is to ensure that the ventricles track the atrial rhythm. The first atrial tachyarrhythmia detection algorithms were developed for bradycardia pacemakers to alleviate patient symptoms that can be caused by inappropriately high ventricular pacing rates which occur when the pacemaker tracks the high atrial rates during AT/AF. This function is called pacemaker modeswitching. Algorithms are designed to recognize the presence of AT/AF and switch the pacing mode to a nonatrial tracking mode during AT/AF, and then switch the mode back to a more physiologic atrial tracking mode when the AT/AF terminates. More sophisticated AT/AF detection algorithms designed for high specificity are implemented into pacemakers, ICDs, and CRT devices that incorporate automatic atrial antiarrhythmic therapies such as atrial antitachycardia pacing (ATP) and atrial cardioversion shocks. High detection specificity is required to minimize potential proarrhythmia and painful, unnecessary shocks. In these systems, rapid detection is not important since AT/AF are not immediately life-threatening and may terminate spontaneously without application of therapy. Faster acting algorithms for discrimination of AT/AF with rapid ventricular response from true ventricular tachyarrhythmias are incorporated into devices such as ICDs that treat life-threatening ventricular tachyarrhythmias (VTs).

The identification of atrial tachyarrhythmias in implantable devices consists of two processes:

1. *sensing* measures the timing of electrical activations in the heart, and

Table 8.1: Clinical application of sensing and detection algorithms.

Implanted device family	Clinical application of algorithm			
	AT/AF monitoring	Timing of pacing stimuli	AT/AF decisions	VT/VF therapy decisions
Subcutaneous monitor	X			
Pacemaker	X	X	X	
Implanted cardioverter/defibrillator	X	X	X	X

Table 8.2: Sensing and detection algorithm functions.

Algorithm	Algorithm function	Clinical application of algorithm			
		AT/AF monitoring	Timing of pacing stimuli	AT/AF decisions	VT/VF therapy decisions
Atrial & ventricular sensing	To appropriately time pacing and defibrillation stimuli.	X	X	X	X
Pacemaker mode switching	To avoid inappropriate ventricular pacing during AT/AF.		X		
Sustained AT/AF detection	To control electrical therapy to terminate AT/AF. To provide diagnostic data for medical management of AT/AF.	X		X	
AT/AF versus VT/VF detection	To discriminate between AT/AF and with rapid ventricular response and true ventricular tachyarrhythmias.	X		X	X

2. *detection* analyzes the timing and morphology of the cardiac signals to determine the presence or absence of a cardiac arrhythmia.

Sensing is achieved by circuits that process the EGM to determine the timing of cardiac depolarizations. Dual chamber devices have separate sensing systems connected to leads implanted in the atrium and ventricle. Detection is achieved by algorithms that process the cardiac timing information with or without the signal morphology information to classify the rhythm. This classification is used to control beat-by-beat paced events, to change the pacing mode, to store data regarding tachyarrhythmias, and to terminate sustained tachyarrhythmias with antitachycardia pacing or shocks.

8.2.1   INTRACARDIAC ELECTROGRAMS

Surface Electrocardiogram versus Intracardiac Electrogram
The ECG is recorded from two electrodes on the body surface some distance from the heart. The locations of both electrodes determine the vectorial "viewpoint" from which the electrical activity of the heart is observed. The amplitude of ventricular activation (QRS complex) is typically around 1 mV and the amplitude of atrial activation is typically less than 0.25 mV.

In contrast, the EGM is a recording from two electrodes, at least one of which is implanted in the heart. The ECG records electrical activity from the entire heart, whereas the EGM accentuates the local wavefronts of depolarization and repolarization that pass the electrode at the tip of the lead. Intracardiac EGMs can be measured from the atrium or ventricle as a unipolar signal (one electrode implanted in the heart and the second electrode is the metal housing of the pacemaker or ICD, e.g., tip to can) or as a bipolar recording (both electrodes residing on the same intravenous pacemaker/ICD lead, e.g., tip to ring). The amplitudes of cardiac activations measured from intracardiac EGMs are typically 10 times greater than the amplitudes seen on the surface ECG. In unipolar systems, the electrode outside the heart may record noncardiac electrical potentials such as from the pectoral muscle. Since unipolar electrode systems are more likely to sense noncardiac potentials, they are contraindicated for ICDs and used infrequently for modern pacemakers, especially when reliable sensing and detection of atrial tachyarrhythmias is desired.

Properties of Atrial EGMs
The amplitude, frequency content, and morphology of EGMs measured in the atrium and in the ventricle depend on a variety of conditions, including the location of the electrodes relative to site of cardiac activation, cardiac rhythm, posture, respiration, and drugs. The amplitude of the atrial EGM during AF tends to be smaller and/or more variable than the amplitude during NSR or other atrial tachyarrhythmias [20]. Atrial EGMs during AF are characterized by extreme temporal and spatial variability. They tend to be most organized when recorded from the right atrial appendage and more disorganized when recorded from other parts of the atrium [20, 21, 22]. As shown in Figure 8.3, visual inspection reveals that the amplitude, width, and morphology of atrial EGMs during AF varies markedly at various anatomic locations [23]. In addition, the spectral coherence of EGMs

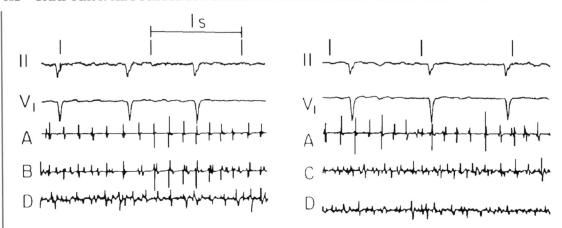

Figure 8.3: Simultaneous recordings of atrial EGMs during AF from different anatomic locations and bipole spacings. A single catheter was placed on the lateral wall in the right atrium. Tracings on the left and right are at two different time epochs. Closer bipoles involving the distal tip electrode on the atrial wall (tracing A) result in recording of more discrete activations during AF, while a wider bipole using the same distal tip electrode (tracing B) result in an EGM that is more disorganized. Tracings C and D also show more disorganization in recordings with two electrodes at least 24 mm distal from the catheter tip. II = ECG lead II, V1 = ECG lead V1, A = Atrial EGM recorded from distal tip to proximal electrode, 1 mm spacing; B = Atrial EGM recorded from distal tip to proximal electrode, 10 mm spacing; C = Atrial EGM recorded from electrodes 24 mm proximal to catheter tip, 10 mm spacing; D = Atrial EGM recorded from electrodes 24 mm proximal to catheter tip, 1 mm spacing. (Reprinted from [23] with permission.)

from two separate atrial sites is greatly reduced during AF as opposed to other atrial rhythms [24]. The effects of recording location and electrode spacing on the characteristics of the atrial EGM can dramatically affect the performance of AT/AF sensing and detection/discrimination algorithms. Antiarrhythmic drugs may also interfere with sensing and detection of AT/AF, as they can cause changes in atrial rate, median frequency, and atrial EGM amplitude [25].

8.2.2   SENSING OF INTRACARDIAC EGMS FOR AT/AF DETECTION

Foundational Components

The goal of the sensing system in a pacemaker or ICD is to determine the timing of cardiac depolarization using EGM signals. Proper operation of tachyarrhythmia detection and pacemaker timing algorithms requires reliable sensing. Sensing in dual chamber pacemakers and ICDs is accomplished by designs that integrate two single chamber sensing systems, one for the atrial and one for the ventricular EGM signal. Each single chamber sensing system uses amplifiers, bandpass filters, signal rectifiers, and comparators to identify atrial or ventricular depolarizations [26]. As shown in Figure 8.4, the foundational components of a pacemaker or ICD sensing system include amplifi-

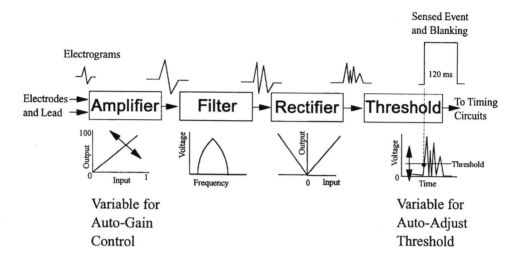

Figure 8.4: Functional block diagram for a pacemaker or ICD sense amplifier. The EGM signal from the two implanted electrodes is first amplified for subsequent processing. Bandpass filtering reduces the amplitude of lower frequency signals such as T waves and far-field R waves and higher frequency signals such as myopotentials and electromagnetic interference. The signal is rectified to make sensing results independent of its polarity. The thresholding operation compares the amplified/filtered/rectified signal to the sensing threshold voltage. At the instant the processed signal exceeds the sensing threshold voltage, the sense amplifier is blanked for a short time period, and a sensed event is declared to pacemaker or ICD timing circuits. For pacemakers, the programmed sensitivity controls the constant sensing threshold voltage. For ICDs, the amplifier gain may be controlled by the input EGM amplitude. The programmable sensing threshold for ICDs controls the high and low limits on the sensing threshold that automatically adjusts on a beat-by-beat basis. (Reprinted from [27] with permission.)

cation, filtering, rectification, and thresholding. The EGM signal passes from the implanted leads into the pacemaker or ICD housing through hermetic feedthroughs with lowpass filters to remove high-frequency electromagnetic interference. After amplification, the EGM signal is processed by a bandpass filter to attenuate extracardiac signals such as myopotential noise and electromagnetic interference and nontarget cardiac signals such as repolarization signals and far-field signals. Far-field signals are EGM deflections that are not from local cardiac tissue, e.g., far-field R waves are frequently observed on an atrial bipolar EGM. Filters typically reject signals with frequencies lower than about 10 Hz and greater than about 60 Hz. The wide range of amplitudes and frequency content of intracardiac depolarization signals poses a challenge for selection of a "perfect" bandpass filter. After amplification and bandpass filtering, the EGM signal is rectified to eliminate the effects of signal polarity and then compared to a sensing threshold. The circuitry declares the presence of a cardiac depolarization event when the processed EGM signal voltage exceeds the sensing threshold. The sensing circuit is "blanked" (turned off) for a period (20–250 ms) after each depolarization is identified to prevent multiple sensed events from being recorded during a single depolarization. Some devices also utilize refractory periods after sensed events that allow recording of sensed events, but help to avoid inappropriate changes in pacemaker timing cycles in response to either oversensing of noise on the EGM signal or proper sensing of tachyarrhythmias.

Sensing Thresholds
Sensing thresholds can be programmed to a constant value (typical in most pacemakers) or automatically adjusted based on the amplitude of each sensed depolarization (typical in most ICDs). Implantable cardiac defibrillators must reliably sense low amplitude ventricular fibrillation in order to detect its presence and deliver life-saving therapy. Automatic adjustment of sensitivity is used with higher sensitivity settings in ICDs to increase the likelihood of sensing low amplitude and varying EGMs during fibrillation, while minimizing the likelihood of oversensing T waves and far-field signals. Ventricular sensitivities are typically around 2.5 mV in pacemakers; ICDs with automatic adjusting sensitivities are programmed with minimum threshold levels of approximately 0.3 mV to ensure sensing of low amplitude ventricular fibrillation. Atrial sensing thresholds are typically 0.3–0.6 mV in pacemakers and ICDs to allow sensing of small amplitude atrial EGMs during AT/AF [28] and to improve the accuracy of atrial tachyarrhythmia diagnostics.

Blanking and Refractory Periods
Blanking periods and refractory periods are used to prevent undesirable behavior caused by oversensing or double counting of cardiac activity. Same-chamber blanking/refractory periods after sensed events reduce double counting of intrinsic cardiac depolarizations. After paced events, the same-chamber blanking/refractory periods are typically longer and prevent oversensing of the pacing artifact and the evoked response. The blanking/refractory periods in the ventricle after atrial sensed or paced events and in the atrium after ventricular sensed or paced events are called cross-chamber blanking/refractory periods. Cross-chamber blanking periods help to prevent oversensing of the pacing artifact after a paced event in the opposite chamber. The atrial blanking period after ventricular events is designed to avoid oversensing of ventricular pacing stimuli and far-field R waves. Longer postventricular atrial refractory periods prevent oversensing of retrogradely conducted atrial activation that may follow ventricular activation or ventricular premature depolarizations.

   Blanking periods may adaptively extend based on noise sampling windows (30–60 ms) if supra-threshold activity (due to cardiac or extracardiac sources such as electromagnetic interference) is seen on the EGM immediately following a sensed event. If noise is seen in consecutive windows following a sensed event, the blanking period is "retriggered" for that beat to avoid double counting or continuous oversensing. This operation has been reported to result in paradoxical undersensing, especially for atrial sensing channels, when more sensitive sensing levels are programmed [29, 30].

Sensing, Blanking, and Refractory Considerations for AT/AF Detection

Atrial tachyarrhythmia detection algorithms rely on accurate estimation of atrial rate. Long atrial cross-blanking periods preclude reliable sensing of most atrial rhythms, especially atrial tachycardias. However, short atrial cross-blanking periods may result in atrial sensing of far-field R waves. Rapidly conducted atrial tachyarrhythmias present the biggest challenge for reliable sensing since the blanked fraction of the cardiac cycle increases as the ventricular rate increases. As a result, pacemakers and ICDs that incorporate atrial tachyarrhythmia detection typically have shorter blanking and refractory periods so that short cardiac cycles can be sensed reliably; see Figure 8.5. Devices that provide CRT by pacing both the right and left ventricles have blanking and refractory periods that are

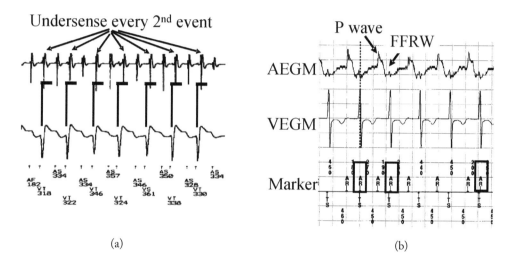

(a)   (b)

Figure 8.5: Reliable sensing of atrial tachyarrhythmias requires short cross-chamber blanking periods that may allow more far-field R wave oversensing. (a) Atrial undersensing during 2:1 atrial flutter due to 86 ms of atrial blanking after each ventricular event. (b) Intermittent oversensing of far-field R waves on the atrial channel in a device with no atrial blanking after ventricular events. (Reprinted from [31] with permission.)

similar to typical dual chamber pacemakers or ICDs. For CRT devices that allow for programmable time delay between right and left ventricular pacing, there are additional blanking periods present for each paced event. The impact of the additional atrial blanking and refractory periods (typically less than 50–100 ms) may impact both atrial and ventricular sensing. This is a special concern for sensing of atrial tachyarrhythmias during high ventricular rates since a larger fraction of the cardiac cycle will be subject to atrial blanking than in non-CRT devices.

It has been demonstrated that as bipolar electrode spacing decreases, the size of the far-field R waves relative to the local P waves decreases [32]. Far-field R wave oversensing on the atrial EGM can be minimized by selection of an atrial lead with a closely spaced bipolar electrode pair (≤10 mm), choosing an implantation location that yields a far-field R wave to P wave ratio of <0.5 [33], or by modifying the programmed sensitivity to reject far-field R waves without undersensing P waves and low amplitude AF.

Some devices may incorporate specialized blanking/sensitivity adjustments after ventricular events that utilize brief atrial blanking periods of reduced atrial sensitivity to reject far-field R waves without preventing detection of AF. Algorithmic rejection of far-field R waves using the pattern of atrial and ventricular events can also be effective, but intermittent sensing of far-field R waves or frequent premature atrial events may disrupt this pattern, resulting in misclassification of a tachycardia. Thus, it is preferable to reject R waves after sensed ventricular events by decreasing atrial sensitivity if this can be done without undersensing AF.

8.3   DETECTION OF ATRIAL TACHYARRHYTHMIAS BY PACEMAKERS, ICDS, AND CRT DEVICES

8.3.1   DETECTION OF AT/AF FOR PACEMAKER MODESWITCHING

Pacemakers programmed to dual chamber synchronous pacing modes may utilize automatic algorithms to initiate a temporary mode change to a nontracking pacing mode during paroxysmal atrial tachyarrhythmias to avoid inappropriately high ventricular pacing rates. There have been several different methods for AT/AF detection for the purposes of modeswitching used by different manufacturers throughout the years. As shown in Table 8.3 all of these methods detect AT/AF when some estimate of atrial rate exceeds a predefined rate threshold. In order to avoid patient symptoms due to rapid ventricular pacing caused by the pacemaker tracking high atrial rates, the time to detection of AT/AF for pacemaker modeswitching should be relatively short (<10 s). Accurate atrial sensing is critical to appropriate mode switching, since all methods depend on measurement of atrial rate and/or A:V patterns. The atrial sensing configuration (unipolar versus bipolar), programmed atrial sensitivity, and atrial blanking periods influence the methods used by different manufacturers and their performance for detection of atrial tachyarrhythmias. Algorithms to recognize repetitive blanking of atrial events during atrial flutter allow mode switches to occur more rapidly when this condition is confirmed.

No matter what method is employed, poor atrial sensing will degrade atrial tachyarrhythmia detection performance [34, 35, 36, 37]. A study of 40 patients who had bipolar pacemakers with Holter monitors demonstrated that mode switching diagnostics appropriately identified 53/54 (98.1%) true atrial tachyarrhythmias with only one short (13 s) false mode switch episode [38]. Since performance is very dependent on the device algorithm, other studies have reviewed stored pacemaker EGMs and have found high percentages of inappropriate mode switching due to atrial oversensing [39, 40]. Care must be taken when relying on mode switching diagnostics before concluding that atrial tachyarrhythmias actually are occurring. Devices with long postventricular atrial blanking periods or low atrial sensitivity may fail to mode switch if a substantial fraction of atrial events are undersensed. Oversensing of far-field R waves or electromagnetic interference/noise in unipolar atrial sensing systems may also cause false positive mode switching.

Table 8.3: Summary of modeswitching algorithms (adapted from [34]).

Algorithm type	Algorithm definition
Mean atrial rate	Identifies atrial tachyarrhythmia when mean atrial rate exceeds programmed threshold. Estimate of mean atrial rate increased by 8–25 ms on short atrial intervals and decreased by 23–39 ms on long atrial intervals.
"Rate and count"	Identifies atrial tachyarrhythmias when a preset number of atrial beats (5–11 beats, sometimes programmable) exceeds a programmable rate threshold. May or may not require confirmation for "X" beats for mode switch to occur.
"X out of Y"	Identifies atrial tachyarrhythmias when X out of Y atrial beats have rate faster than a rate threshold or have short ventriculo-atrial intervals.
Beat-to-beat	Switches mode based on single short atrial interval or short ventriculo-atrial interval.
Combined Algorithms	Combination of Rate and Count or X out of Y and beat-to-beat for special cases.

8.3.2 DETECTION OF ATRIAL TACHYARRHYTHMIAS FOR ATRIAL ANTIARRHYTHMIC THERAPY

Detection of atrial tachyarrhythmias in a device designed for atrial antiarrhythmic therapy must be performed with high specificity to avoid over-treatment of nonsustained atrial tachyarrhythmias. The desired time course for automatic therapy for atrial tachyarrhythmias may vary from several minutes to several hours, depending on patient symptoms and acceptance of the therapies. This is in contrast to automatic detection and therapy for VTs which must be performed with high sensitivity and within several seconds to avoid loss of patient consciousness and arrhythmic death.

High specificity for AT/AF detection has been achieved by use of more sophisticated algorithms which incorporate atrial rate estimates with pattern based algorithms to recognize atrial tachyarrhythmias with higher atrial rate than ventricular rate (e.g., >1:1 A:V pattern) and reject far-field R wave oversensing on the atrial sensing channel. These algorithms also include methods for discrimination of organized AT from AF or disorganized AT. Algorithms for AT/AF discrimination are important for guiding therapies, since AT rhythms are more likely to respond to atrial antitachycardia pacing. Two current approaches that achieve AT/AF detection and discrimination are highlighted in the sections below; other algorithms for identification of rapid atrial rates or when atrial EGMs with characteristics of AF are present have been previously described, but are not currently in use clinically [26, 41, 42, 43].

A:V Pattern and Regularity Based AT/AF Detection

The AT/AF detection algorithm in the Medtronic Jewel AF, EnTrust DR, Concerto DR ICDs, and in the EnRhythm pacemakers achieves high specificity by combining dual chamber (atrial and ventricular, or P:R) pattern information and median atrial rate (interval) thresholds to detect and classify non-1:1 atrial tachyarrhythmias. The 7th longest of the last 12 atrial intervals is used to estimate the median atrial interval in order to provide some immunity to atrial undersensing or oversensing. On each ventricular event, the median atrial interval is calculated and the A:V pattern is assessed; see Figure 8.6. The AT/AF evidence counter accumulates A:V pattern information using an up–down counting scheme (minimum value of 0, maximum value of detection threshold +15). The counter increments by one on each ventricular event where there are two or more atrial events and there is no pattern-based evidence of far-field R wave oversensing on the atrial channel. The counter decrements by one if there is strong evidence of lack of N:1 rhythm (e.g., two consecutive 1:1 beats). Isolated 1:1 beats contribute AT/AF evidence if they are preceded by a confirmed N:1 beat. The evidence counter has two stages of operation. The first stage defines the start of an AT/AF episode when the evidence counter reaches a predefined threshold (either 24 or 32 depending on the specific device). Once the start of an episode is defined, the episode duration timer begins incrementing and the evidence counter switches to the sustained detection stage. To allow for intermittent atrial undersensing during this stage, the evidence counter accumulates an additional 15 "counts" beyond the preliminary detection threshold, and the AT/AF pattern criterion remains satisfied as long as the evidence counter is greater than 16 or 24 (depending on the specific device). After atrial antitachycardia therapies are delivered, the AT/AF evidence counter is reset to zero, and an episode is redetected when the evidence counter becomes satisfied again.

A measure of atrial interval regularity is computed to aid in the discrimination of atrial rhythm transitions from AT to AF, and vice versa. The atrial interval regularity criterion is evaluated on each ventricular event and is based on the most recent 12 atrial intervals. Each set of 12 atrial intervals is considered "regular" if the difference between the shortest and longest atrial cycle length is no more than a percentage of the median atrial cycle length (nominally 25%). Rhythms are considered "regular" when 12 of the last 12 overlapping sets of atrial intervals are deemed to be regular. The A:V pattern information is also used to recognize termination of atrial tachyarrhythmias, defined as five consecutive ventricular events with a NSR A:V pattern. The criteria for NSR PR pattern uses strict PR timing restrictions to guard against undersensing conditions that could falsely trigger recognition of episode termination.

Recognition of far-field R wave oversensing on the atrial channel is based on PR pattern and timing information. Far-field R wave oversensing during sinus rhythm or sinus tachycardia is identified when there are exactly two P waves for each RR interval, alternating PP intervals in a short-long pattern (>30 ms difference), low RP interval variability (<50 ms), and consistent RP intervals or PR intervals (<20 ms from the average RP or PR). The consistent alternation of PP intervals is needed to avoid misclassifying the rhythm as 2:1 atrial flutter. One of the P events must be close to the R event (PR<60 ms or RP<160 ms).

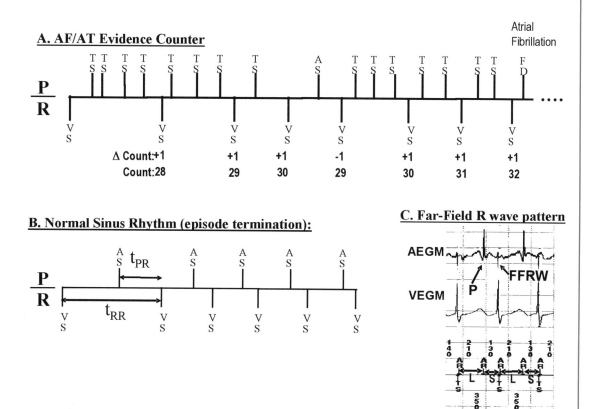

Figure 8.6: Patterns used in AT/AF detection. The stick diagrams represent atrial sensed events (markers above the horizontal line) and ventricular events (markers below the horizontal line) used to generate the pattern information. (a) The AT/AF evidence counter is evaluated on each ventricular event. The counter increments when there are two or more atrial events between ventricular events and when there is less than two atrial events and the evidence counter incremented on the prior event; otherwise, the counter decrements. (b) Presence of NSR pattern indicates AT/AF episode termination. The NSR pattern requires five consecutive beats of 1:1 rhythm with specific P:R timing restrictions as indicated. Normal sinus rhythm can also be recognized during 2:1 rhythms with far-field R wave oversensing pattern. (c) The far-field R wave oversensing pattern requires exactly two atrial events between ventricular events. "S" and "L" indicates a short–long alternating pattern of atrial cycle length, one of four specific criteria used to exclude atrial tachyarrhythmias with 2:1 conduction from being classified as far-field R waves.

The performance of this rate and pattern-based AT/AF detection algorithm was prospectively evaluated in 58 patients, using a custom telemetry Holter device to simultaneously record ECG and device markers for 24 h. The results of this analysis validated the continuous detection of AT/AF with a sensitivity of 100% and a specificity of 99.9% (116 h of AT/AF, 1290 hours of non-AT/AF) [44].

In newer Medtronic devices, the "reactive ATP" algorithm resets the antitachycardia pacing timer to permit repeated attempts at antitachycardia pacing when changes in atrial rate or atrial rhythm regularity are detected, or after a preprogrammed duration of sustained AT/AF. The premise of this approach is that changes in rate or regularity of the atrial intervals represent a change to a different AT/AF rhythm which may be more susceptible to ATP termination. The therapy scheduling scheme divides atrial rhythms into bins based on the median atrial interval and atrial interval regularity as described above; see Figure 8.7. Figure 8.8 presents an example of successfully terminated AT/AF where the reactive ATP therapy scheduling algorithm recognized the presence of a slower, regular AT/AF several hours after a prior unsuccessful therapy for a faster AT/AF rhythm.

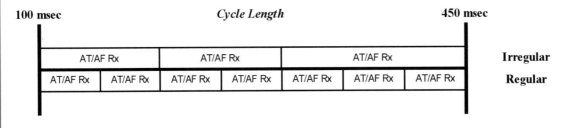

Figure 8.7: Reactive ATP scheduling algorithm divides all AT/AF episodes into 10 bins based on median atrial interval and regularity of atrial intervals. Each bin is assigned a specific and finite set of ATP therapies that are delivered according to the detection result. Once an AT/AF rhythm has been detected, the atrial intervals are analyzed and the rhythm is assigned to one of the bins and the first programmed therapy is delivered. If the therapy is unsuccessful, the episode is redetected and the process is repeated. After the set of therapies for any bin have been depleted, no additional therapies can be delivered until the rhythm changes and is reclassified into a new bin.

AT/AF Detection using Atrial Intervals Alone

The atrial rhythm classification (ARC) algorithm in the Guidant Ventak Prizm AVT implantable atrial and ventricular defibrillator detects the presence of atrial tachyarrhythmias when 32 of the last 40 atrial intervals are faster than a programmed rate cutoff. The ARC algorithm discriminates between AF and AT (organized atrial tachyarrhythmias such as atrial flutter) based on characterizing the 12 most recent atrial intervals in terms of maximum rate (shortest interval) and two measures of atrial interval variability [45]. The range of the atrial intervals is one measure of variability that captures the difference between the longest and shortest A–A intervals. The second measure of

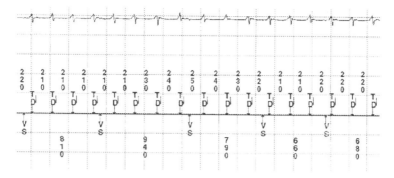

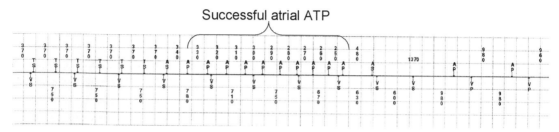

Figure 8.8: Example of successfully terminated AT/AF. The top panel shows the atrial EGM and marker diagram from an AT/AF episode (median atrial interval 220 ms) that was initially detected and treated unsuccessfully with ATP therapy. After 2 h, the atrial rate slowed, and the reactive ATP algorithm recognized the presence of a slower, regular AT/AF rhythm with a median atrial interval of 370 ms. In the lower panel, atrial ATP was delivered and successfully terminated the AT/AF. TD = AT/AF detected, VS = ventricular sense, TS = AT/AF sensed event, AP = atrial pace, VP = ventricular pace.

atrial interval variability is the standard deviation of A–A intervals. Values of these three variables define a point in a three-dimensional space. A curved surface separates the AF region from the atrial flutter region. Points in the AF region have higher atrial rate, higher range of A–A intervals, and greater standard deviation of A–A intervals. The ARC continuously evaluates the AT/AF rhythm classification decision using a sliding window of 12 consecutive atrial intervals until the end of the episode is reached. An additional "X of Y" rhythm classification criterion can be used to filter transient errors in beat-to-beat AT/AF classification results. For example, if at least 8 out of the last 12 discrimination decisions return the result of AF, then the classification result for the rhythm will be AF (otherwise the rhythm decision will be AT) [45, 46]. A prospective evaluation of AT/AF detection and discrimination using the ARC algorithm showed a higher incidence of ATP therapy success for rhythms classified as AT compared to rhythms classified as AF (67% versus 26%) [46]. The higher ATP success incidence for AT classified rhythms provides some positive evidence for the appropriateness of the AT/AF discrimination metrics used by this algorithm.

8.3.3   DETECTION OF AT FOR VT/SVT DISCRIMINATION IN ICDS

Ventricular rates during supraventricular tachycardias (SVTs), including AT/AF, can overlap with the ventricular rates observed during VTs (typically >150 bpm). In order to increase the accuracy of VT detection, ICDs and some pacemakers with advanced ventricular arrhythmia diagnostics utilize algorithms for discrimination of VT and SVT with rapid ventricular rate. These algorithms include specific means for discriminating rapidly conducted AT/AF from VT and are designed to react in a time course commensurate with the time course for device-delivered ventricular arrhythmia therapy (3–6 s). This is in contrast to AT/AF detection algorithms that control atrial ATP and cardioversion therapies, which are typically 20 s or longer. Devices that provide electrical therapies for both VT/VF and for AT/AF (e.g., PRIZM AVT from Guidant; EnTrust AT and Concerto AT from Medtronic, Inc) contain both types of AT/AF detection algorithms arranged in a hierarchy, with the VT/SVT discrimination algorithms taking priority over the AT/AF therapy controlling algorithms when the ventricular rate exceeds the rate cutoff for VT detection.

Devices designed to detect VT analyze ventricular intervals to determine the presence or absence of a tachycardia that is fast enough for a sufficiently long duration. The interval analysis methods are typically simple counting methods that accumulate the number of intervals shorter than a programmable threshold (beats with instantaneous rate faster than a rate threshold). Once the counter exceeds a predetermined threshold, a tachycardia is declared. The ventricular rate during AF is generally characterized as being irregular, with both long (slow) and short (fast) intervals during a period of high average heart rate, while the ventricular rate during VT is generally very regular. As a result, interval analysis methods that require consecutive intervals to be faster than the rate threshold provide some additional VT/AF discrimination power compared to other methods [32].

Discrimination of rapidly conducted AT/AF from VT can also be accomplished by explicit measures of ventricular rate regularity or by ventricular EGM morphology information. Ventricular rate regularity methods may lose effectiveness with faster average ventricular rates during AF (when AF tends to conduct more regularly) [47] and may be ineffective for rhythms with regular ventricular response, such as 2:1 atrial flutter. Electrogram morphology analysis methods are unaffected by ventricular rate regularity and can discriminate AT/AF from VT on the premise that ventricular depolarizations (QRS complexes) during AT/AF are similar to those during NSR, since the origin of the electrical activation causing the depolarization is in the atrium for both rhythms. This is in contrast to the QRS complexes during VT which are typically different than those during NSR since the origin of the electrical impulse is in one of the ventricles. Discrimination of SVT, including AT/AF with regular and irregular ventricular rates, and VT is achieved by analyzing the EGM waveform and identifying a rhythm as SVT if the EGM characteristics during tachycardia are similar to the EGM characteristics during beats of known atrial origin [48, 49]. These methods may be confounded by abnormal conduction of atrial impulses to the ventricle during rapid AT/AF (i.e., aberrant conduction) and VT with EGM morphology that is similar to the patient's intrinsic rhythm when viewed from the single EGM vector used for analysis [50, 51].

Dual chamber and CRT devices incorporate the AF/VT discriminators described above, but also use information from the atrial EGM to improve the specificity and sensitivity of VT detection. Explicit or implicit calculation of atrial rate relative to ventricular rate, measures of A:V association, and A:V patterns derived from the atrial timing measured relative to the ventricular timing can provide additional specificity for regularly conducted AT/AF and for AT/AF with aberrant conduction by helping to confirm that the ventricular rate is being driven by the electrical impulses in the atria. Sensitivity for VT detection is improved by applying the ventricular regularity or EGM morphology discriminators only when the atrial rate is confirmed to be equal to or higher than the ventricular rate, eliminating false negative detection of obvious VT (when the ventricular rate is faster than the atrial rate). Several different SVT/VT discrimination algorithms that incorporate atrial information have been developed, each with different detailed implementations but using similar building blocks of information derived from the EGM signals as described above [52, 53]. Adequate atrial sensing is key for achieving improved SVT/VT discrimination performance with dual chamber algorithms [54]. Despite the significant improvements that have been made in SVT/VT discrimination algorithms over the past decade, inappropriate shocks for rapidly conducted SVTs (mostly AF), self-terminating VT, and ventricular oversensing remain a key clinical challenge and an important driver for further improvements to device-based discrimination algorithms.

8.4   ATRIAL TACHYARRHYTHMIA DETECTION WITH SUBCUTANEOUS MONITORING DEVICES

Unlike pacemakers and defibrillators, monitoring devices typically do not have sensing leads in the heart and hence a different approach for atrial tachyarrhythmia detection is adopted. External monitoring devices are presently being utilized to store automatically triggered AT/AF episodes [8, 10, 14, 55]. However, these commercially available cardiac event recorders do not have computational constraints as required for implantable devices. One AF detector used a combination of the wavelet transform and the power spectrum to discriminate between AF and NSR [55]. Another AF detector used variability of RR intervals and the morphology of the QRS complexes to detect AF [10]. Use of autocorrelation, power spectrum [56], coefficient of variation, and comparison of distributions [57] to differentiate AF from NSR has also been reported in the literature. Lorenz plots of RR intervals have also been used to visualize RR interval patterns during AF [58, 59, 60]. None of these publications have addressed the problem of AT detection. Atrial tachycardia, which includes atrial flutter, is as clinically relevant as AF, as both arrhythmias increase the risk for stroke in the presence of other co-morbidities. Therefore, it is important to provide AT detection capabilities in a monitoring device that provides diagnostics for atrial arrhythmias. Since the ventricular response during AT may be different than what is observed during AF, most of the external AT/AF detectors will not be able to detect AT accurately. A device with a detector that can measure AT/AF burden accurately could provide daily monitoring of AT/AF over a long period of time and impact therapeutic decisions.

Presently, AT/AF patients who are not indicated to receive a pacemaker or ICD cannot be monitored continuously over a long duration of time. Automatic long-term AT/AF monitoring could be provided by chronically implanted loop recorders (ILR) with closely spaced electrodes on the surface of the device. Such recorders are typically implanted subcutaneously for recording ECG during cardiac events [61]. Asymptomatic brady- and tachyarrhythmia events can be automatically detected and ECG storage for symptomatic events can be initiated by the patient with the help of a handheld activator. Unlike pacemakers and ICDs, in which AT/AF is detected using information from intracardiac sensing leads in the atrium and the ventricle, ILRs would need to detect AT/AF based on RR intervals alone, as the P wave amplitudes tend to be relatively small. Due to the limited processing power and constraints of maximizing battery life of the device, the AT/AF detection algorithms should be based on RR intervals and computationally simple.

8.4.1   SENSING OF RR INTERVALS

Robust R wave sensing is a first step for detection of AT/AF in an ILR. The lower amplitude of R wave signals and the higher possibility of noise and artifact make R wave sensing for a subcutaneous device more challenging than with devices utilizing intracardiac signals as described in the previous section. A typical ECG recording from an ILR device over a period of 30 s is shown in Figure 8.9. This ECG was measured from a subcutaneous dipole with an interelectrode spacing of 4 cm and an

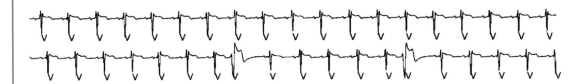

Figure 8.9: An ILR-stored ECG strip from a patient with premature ventricular beats. Each sensed R wave is marked with a "V."

electrode surface area of 42 mm². The ILR ECG signal characteristics are different than a near-field signal with bipolar electrodes within the heart. The far-field nature of the measurement makes an ILR ECG similar to a surface ECG measurement with lower frequency content than a near-field measurement. The amplitude of R waves recorded by an ILR averages around 300 μV compared to amplitudes on the order of 1–2 mV recorded in the precordial leads of a surface ECG. R wave sensing in an ILR uses the same principles as shown in Figure 8.4 with modifications to the characteristics of the bandpass filter, the blanking period, and noise rejection algorithms. For sensing of R waves, bandpass filtering from 10 to 32 Hz with notch filters at 50 and 60 Hz are typically utilized. The bandpass frequency response is similar to that used for intracardiac EGMs with the different cutoff frequencies. The refractory periods typically extend to about 130 ms following the R wave and may even extend further if the sensing system detects noise.

8.4.2   PHYSIOLOGICAL CONCEPTS FOR DESIGN OF DETECTION ALGORITHM

Ventricular response during AF is irregular or random in nature [56, 62]. Multiple atrial impulses continuously arrive at the AV node in a random fashion, leading to various degrees of concealed conduction through the AV node depending upon its refractory state [62, 63, 64]. Further, the conductivity and refractoriness of the AV node and atrial tissue may be altered by changes in autonomic tone and pharmacologic agents. Most investigators have reported a very low autocorrelation of RR intervals during AF to establish randomness [56, 62]. Nonlinear forecasting techniques have been used to show that in a significant number of patients, ventricular response during AF is weakly predictable, possibly due to cyclic oscillations in autonomic tone [65]. Spectral analysis showed that RR intervals during AF have a white noise-like spectrum in the short-term (several minutes) and a 1/f characteristic over a longer term (for frequencies below 0.005 Hz [56]). The fundamental premise for the AT/AF detector designed in that work was that the RR intervals during AF are irregular and uncorrelated over several minutes.

The pattern of ventricular response during AT has not been widely reported in the literature. Atrial tachycardia can have regular, irregular, and regularly irregular ventricular responses depending on the response of the AV node to atrial activation. Atrial tachycardia with regular ventricular response is a result of regular atrial activations with a consistent A:V conduction ratio (e.g., 2:1, 3:1, etc.). Atrial tachycardia with irregular ventricular response results from irregular atrial activations and inconsistent AV node conduction for each cycle, as during AF. Atrial tachycardia with regularly irregular ventricular response is a result of the mechanism called "group beating" [66]. In this case, the different degrees of block in the AV node may lead to different ventricular intervals such as short–short–long, short–long–short, long–short–long, or long–short–short RR intervals. An AT/AF detector based on irregularity of ventricular response will only detect AT with irregular RR intervals. A supplemental AT detector is required to detect AT with regular and regularly irregular RR intervals.

8.4.3   LORENZ PLOT DISTRIBUTION OF RR INTERVALS DURING ATRIAL TACHYARRHYTHMIA

The irregularity and regularity of the RR intervals during AT or AF was the fundamental information utilized to design a detector [67]. The δRR interval, defined as the difference between the present and the previous RR intervals,

$$\delta RR(i) = RR(i) - RR(i-1),\tag{8.1}$$

is used as a measure of irregularity. The Lorenz plot of δRR intervals, which is a scatter plot of $\delta RR(i)$ versus the previous $\delta RR(i-1)$, encodes this incremental information in the direction of change of three consecutive RR intervals[1]. Figure 8.10 illustrates how RR intervals, derived from an ECG during AF, populate the Lorenz plot of δRR intervals. Each point $(\delta RR(i), \delta RR(i-1))$ has a magnitude and phase. The magnitude encodes the irregularity and the magnitude coupled

---

[1]It should be noted that this type of plot is also referred to as the Poincaré plot, cf. Chapter 5. In this chapter, this type of plot is based on RR interval differences, whereas it is directly based on the RR intervals in Chapter 5.

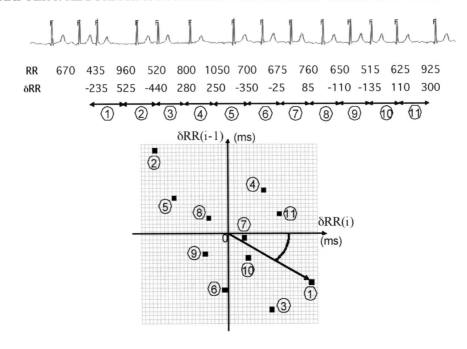

| RR | 670 | 435 | 960 | 520 | 800 | 1050 | 700 | 675 | 760 | 650 | 515 | 625 | 925 |
| δRR | | -235 | 525 | -440 | 280 | 250 | -350 | -25 | 85 | -110 | -135 | 110 | 300 |

Figure 8.10: Illustration showing how RR intervals, derived from an ECG during AF, populate a Lorenz plot of δRR intervals. Each point $(\delta RR(i), \delta RR(i-1))$ in the plot is derived from four R waves which provide three RR intervals. The diagram axes extend from $[-600, 600]$ ms.

with phase encodes the incoherence of changes in RR intervals. Further, it is possible to detect characteristic signatures of the cluster of points in a Lorenz plot during AT/AF. Examples of ECG segments during AF and AT along with the corresponding Lorenz plot of δRR intervals are shown in Figure 8.11. A two-dimensional histogram was created, being a numeric implementation of the Lorenz plot, and divided into multiple segments as shown in Figure 8.12(a). The sequence of RR intervals that populate these segments is tabulated in Figure 8.12(b). For example, segments 1 and 9 have an RR sequence of short–long–short (S–L–S), i.e., a positive $\delta RR(i-1)$ is followed by a negative $\delta RR(i)$. Segment 9 will be populated with points for which $|\delta RR(i)| \approx |\delta RR(i-1)|$. The width of segment 9, which is same as the width of segment 0 defined by the parameter *NSRmask*, accounts for the variation due to autonomic modulation of the AV node. Segment 1, on the other hand, has a similar S–L–S sequence, but $|\delta RR(i)| \neq |\delta RR(i-1)|$. Similarly, the other RR interval sequences in other segments can be explained using Figure 8.12(b).

During AF, due to the lack of correlation of one RR interval to the next, the points are distributed across all segments in a Lorenz plot; see Figure 8.13(f). The distribution has a random behavior and looks similar to a Lorenz plot generated by white noise. The density of the distribution during AF varies with the underlying cycle length since the coefficient of variation of RR intervals

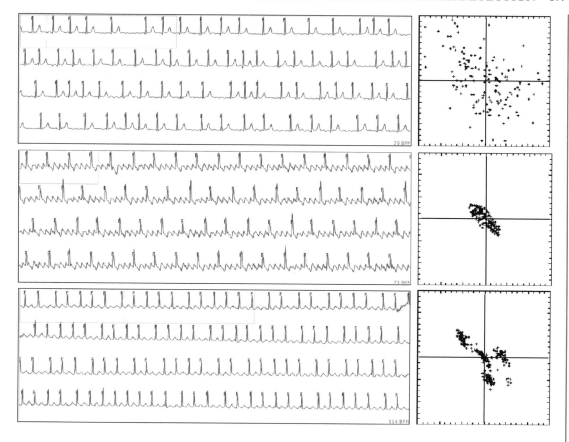

Figure 8.11: Three ECG examples during AT/AF and the corresponding Lorenz plots. A 2-min strip of AF is shown in the uppermost example, and 2-min strips of atrial flutter are shown in the next two examples.

is almost constant during AF [57, 58, 63]. At shorter cycle lengths, the irregularity reduces and the distribution becomes denser, whereas at longer cycle lengths the distribution becomes sparser. Although δRR intervals normalized by rate could be used, we used a computationally simpler approach. The δRR intervals were divided or multiplied by 2 if the cycle length was long or short, respectively, based on a predefined threshold. Lorenz plots of δRR intervals during NSR is shown in Figure 8.13(a). During NSR most points $(\delta RR(i), \delta RR(i-1))$ are within segment 0. There are other non-AF rhythms, specifically a series of premature atrial contractions, that have irregular RR intervals and exhibit specific signatures in the Lorenz plot as shown in Figures 8.13(g)–(i). These non-AT/AF rhythms have specific signatures with a higher probability of a cluster of points

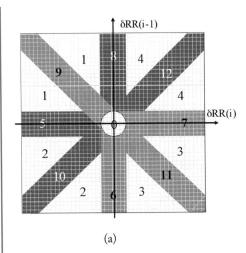

#	Sequence	δRR
0	S–S–S / L–L–L	$|\delta RR| < NSRmask$
1	S–L–S	$|\delta RR(i-1)| \neq |\delta RR(i)|$
2	L–M–S	$|\delta RR(i-1)| \neq |\delta RR(i)|$
3	L–S–L	$|\delta RR(i-1)| \neq |\delta RR(i)|$
4	S–M–L	$|\delta RR(i-1)| \neq |\delta RR(i)|$
5	L–L–S	One $|\delta RR| < NSRmask$
6	L–S–S	One $|\delta RR| < NSRmask$
7	S–S–L	One $|\delta RR| < NSRmask$
8	S–L–L	One $|\delta RR| < NSRmask$
9	S–L–S	$\delta RR(i-1) \approx -\delta RR(i)$
10	L–M–S	$\delta RR(i-1) \approx \delta RR(i)$
11	L–S–L	$\delta RR(i-1) \approx -\delta RR(i)$
12	S–M–L	$\delta RR(i-1) \approx \delta RR(i)$

(a)   (b)

Figure 8.12: (a) The two-dimensional histogram, a numeric representation of a Lorenz plot, of δRR intervals. Thirteen segments are marked on the plot denoting regions that would be populated by points for different sequences of RR intervals as tabulated in (b). *NSRmask* is the radius of segment 0 and has a nominal value of 80 ms. The histogram was divided into segments 1–12 primarily for supplementary AT detection. S, M, and L denote short, medium, and long, respectively.

occurring in specific segments. The detector should only detect the rhythms in Figures 8.13(b)–(f) and not falsely detect during the other rhythms shown in Figures 8.13(g)–(i).

Lorenz plots of δRR intervals during AT with different degrees of organization are shown in Figures 8.13(b)–(e). The most organized form of AT, in which the RR intervals are very regular, is shown in Figure 8.13(b). On the other hand, Figure 8.13(e) shows the distribution of an AT with disorganized ventricular response that is similar to AF in terms of irregularity of RR intervals. Figures 8.13(c)–(d) show different degrees of discrete organization of ventricular response due to varying atrioventricular conduction. Figures 8.13(c)–(d) are from different patients suggesting that this is a very common signature of discrete AV node conduction during AT. The clusters indicate a change in AV conduction ratio, and the variability within each cluster implies varying degrees of autonomic modulation of the AV node. A continuum of organization from Figures 8.13(c)–(e) and all possible conduction ratios with different degrees of AV node modulation are possible. Figure 8.13(e) is a more organized form of AT compared to Figure 8.13(f) because of the compactness of the former distribution. Overall, Figures 8.13(c)–(e) shows that there is a higher probability of points populating segments 6, 7, 9, and 11 during AT.

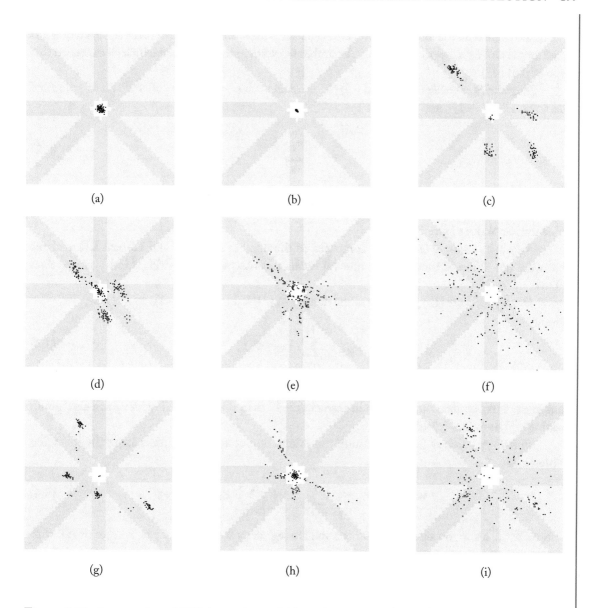

Figure 8.13: Lorenz plot of δRR intervals overlaid on the segmented two-dimensional histogram for 2-min data during (a) NSR, (b) AT with regular ventricular response, (c)–(d) AT with "group beating" with various degrees of organization, (e) AT with irregular ventricular response, (f) AF, and (g)–(i) series of premature atrial contractions with various levels of irregularity in ventricular response (*NSRmask* = 80 ms). The plots extend from −600 to +600 ms along both axes.

8.4.4   ATRIAL TACHYARRHYTHMIA DETECTION ALGORITHM

The detector encodes the information regarding patterns in the distribution of $(\delta RR(i), \delta RR(i-1))$ in a Lorenz plot over a period of 2 min. A number of metrics are introduced which characterize this type of plot [67].

Irregularity evidence (denoted $\mathcal{L}_{IE}$) measures the sparseness of the distribution of $(\delta RR(i), \delta RR(i-1))$, and is defined by

$$\mathcal{L}_{IE} = \sum_{k=1}^{12} B_k \,, \tag{8.2}$$

where B_k denotes the number of histogram bins in segment k of the Lorenz histogram that are populated by at least one point. This metric has a high value during AF (Figure 8.13(f)) and a low value during NSR (Figure 8.13(a)).

Regularity evidence ($\mathcal{L}_{RE}$) was computed as the number of 6-beat and 12-beat RR medians that were not different than the RR interval median of the previous 2-min period by more than 10 ms.

Density evidence ($\mathcal{L}_{DE}$) measures the density of $(\delta RR(i), \delta RR(i-1))$ in a cluster,

$$\mathcal{L}_{DE} = \sum_{k=5}^{12} (P_k - B_k) \,, \tag{8.3}$$

where P_k denotes the number of points that populate segment k. This metric has a high value when multiple points populate the same bin in segments 5–12 as in the case of AT with ventricular "group beating" (Figures 8.13(c)–(d)).

Anisotropy evidence ($\mathcal{L}_{AE}$) measures the orientation of the distribution using the following expression:

$$\mathcal{L}_{AE} = \left| \sum_{k=9,11} P_k - \sum_{k=10,12} P_k \right| + \left| \sum_{k=6,7} P_k - \sum_{k=5,8} P_k \right| . \tag{8.4}$$

This metric has a high value if the points populate segments 6, 7, 9, and 11 as in cases of AT with irregular ventricular response (Figures 8.13(c)–(e)).

Premature atrial contraction evidence ($\mathcal{L}_{PE}$) measures the presence of a dense distribution in segments 1–4 which is a characteristic of compensatory pauses,

$$\mathcal{L}_{PE} = \sum_{k=1}^{4} (P_k - B_k) + \sum_{k=5,6,10} (P_k - B_k) - \sum_{k=7,8,12} (P_k - B_k) . \tag{8.5}$$

This type of pattern is exemplified in Figures 8.13(h)–(i). Additionally, $\mathcal{L}_{PE}$ measures the presence of density in segments 5, 6, and 10, and the corresponding absence in segments 7, 8, and 12 as seen in examples in Figures 8.13(g)–(h).

The metrics *AF evidence* ($\mathcal{L}_{AFE}$), *AT evidence* ($\mathcal{L}_{ATE}$), and *organization index* ($\mathcal{L}_{OI}$) are computed from the distributions and combined to form the "AT/AF mode" or the "supplemental AT detection mode" of the detector. The metrics $\mathcal{L}_{AFE}$ and $\mathcal{L}_{ATE}$ quantify the degree to which the Lorenz plot suggests the presence of AF and AT, respectively, defined by

$$\mathcal{L}_{AFE} = \mathcal{L}_{IE} - \mathcal{O} - 2\mathcal{L}_{PE} \,, \tag{8.6}$$
$$\mathcal{L}_{ATE} = \mathcal{L}_{IE} + \mathcal{L}_{AE} + \mathcal{L}_{DE} + \mathcal{L}_{RE} - 4\mathcal{L}_{PE} \,, \tag{8.7}$$

where $\mathcal{O}$ denotes the number of points in the bin containing the origin. The metric $\mathcal{L}_{OI}$ quantifies the degree of organization in an atrial arrhythmia by evaluating the degree of organization in the ventricular response,

$$\mathcal{L}_{OI} = \mathcal{O} + \mathcal{L}_{RE} + \mathcal{L}_{AE} + \mathcal{L}_{DE} - 2\mathcal{L}_{IE} \,. \tag{8.8}$$

The metrics $\mathcal{L}_{AFE}$, $\mathcal{L}_{ATE}$, and $\mathcal{L}_{OI}$ are used to assert the state of the detector every 2 min as AF, AT, or NSR.

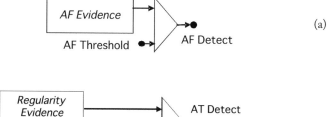

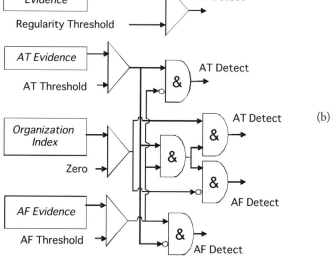

Figure 8.14: The logic used on the accumulated evidence over a 2-min period to assert the state of the detector. (a) AT/AF detector and (b) supplemental AT detector.

The detector operates in the base "AT/AF mode", which detect AF and AT with irregular ventricular response, by comparing $\mathcal{L}_{AFE}$ to a threshold to detect AF, see Figure 8.14(a). In the "supplemental AT mode" (Figure 8.14(b)), AT and AF are detected with a combination of different threshold tests. This mode also incorporates capabilities that can detect AT with "group beating" and regular ventricular response as exemplified in Figures 8.13(b)–(d).

8.4.5   DETECTOR PERFORMANCE

The detector was optimized to measure AF burden [67]. Multiple databases having significant durations of AF, AT, and sinus rhythm were used to evaluate the performance of the detector. The databases consisted of 24-h Holter recordings from 422 patients and included the MIT–BIH AF database [68] which had 10 h of recordings from each patient. The AF detector was found to be accurate in measuring AF burden. Both AF burden sensitivity and specificity, computed by comparing detection and clinical truth in 2-min blocks, were 99% in databases comprised of AF patients and non-AF patients (excluding patients with AT). The AF detector performance reduces if the patient experiences a significant amount of AT. Incorporation of the supplemental AT detector improves the ability to detect AT; however, the specificity of AT/AF burden detection was compromised. The AT/AF detector with supplemental AT detection showed AT/AF burden sensitivity of 92% and specificity of 97%, in datasets with AT, AF, and non-AT/AF.

For patients with AT/AF burden exceeding 10 min, a metric of burden accuracy was defined as the number of patients with detected burden within 20% of the true burden. In patients having only AF burden, the AT/AF detector was found to have good accuracy in 96% of patients. In a mix of 10% AT and 90% AF burden, i.e., a typical patient population, the AT/AF detector was found to have good accuracy in 92% of patients. Atrial fibrillation was overestimated in 3% of patients, while AT was underestimated in 5% of patients. More than 10 min of AF was detected in 6% of the records without AT/AF. Underestimation of AT was reduced 2% by adding the supplementary AT detector; however, the overall accuracy was reduced by 6% due to overestimation of burden in patients with low AT/AF burden.

8.5   MONITORING AND CLINICAL DIAGNOSTICS FOR ATRIAL TACHYARRHYTHMIAS WITH ICDS

After cardiac signals have been sensed and interpreted through detection algorithms, implantable devices must summarize and present this information to physicians and nurses in a way such that clinically relevant decisions can be made. The goals of this section are to discuss how data is collected and organized within implantable devices and to present examples of how device diagnostics are used in both the rhythm control and rate control strategies for patients with AT/AF.

8.5.1   DATA COLLECTION AND ORGANIZATION

While implantable devices are capable of continuously monitoring cardiac rhythms over the life of the device, there are two important reasons why this information must be condensed and summarized.

First, memory limitations within the device would not permit continuous storage of EGM and other data for each cardiac cycle. A patient experiences approximately 100,000 heartbeats per day and may go 6 months between follow-up visits where the data can be extracted from the device via telemetry. This means the device may monitor more than 18 million heartbeats between office visits. Secondly, this amount of data would be far too much for the physician to process and interpret for each patient. The challenge for device manufacturers is to convert this vast amount of detection information into meaningful diagnostic information that is manageable by both the device and the clinician.

To accomplish this, the device stores atrial arrhythmia information in a hierarchical fashion. Extremely detailed information is only stored for a small subset of the episodes, while very general information is tabulated across all episodes. For example, one of the most memory intensive pieces of information is the EGM waveforms. This data can be crucial for the clinician to verify that the device is detecting correctly and aid in troubleshooting when it is not. Because this information requires extensive amounts of memory (as well as additional battery resources), only small portions of EGMs from a select number of episodes are stored by the device. On the other end of the spectrum, a continuous running tally of the duration of all AT/AF episodes is important so the physician can assess the AT/AF burden experienced by the patient. Between these two extremes are a variety of other parameters that may be tabulated per episode, per day, or per follow-up period. The goal is to strike a balance between providing enough information to be clinically useful in managing the patient without providing so much data as to overwhelm the clinician. Specific examples of device diagnostics for rhythm control and rate control of patients with AT/AF will be presented in the remainder of this chapter.

8.5.2   MONITORING FOR RHYTHM CONTROL

As mentioned earlier, rhythm control is one of the two primary strategies for treating patients with AT/AF. The objective of this strategy is to maintain the patient in sinus rhythm. Therefore, one of the most important diagnostics for rhythm control is to know what percentage of time was spent by the patient in an atrial tachyarrhythmia, i.e., AT/AF burden. Accurate monitoring is crucial regardless if the patient is being treated by drugs, ablation, or device therapies. Figure 8.15 shows a series of patients who were treated for AT/AF by undergoing a pulmonary vein ablation procedure at month 0. These patients had a device implanted for their bradycardia which was also used to monitor the effectiveness of the ablation procedure. The amount of AT/AF experienced each day is plotted over the follow-up period. This example also illustrates that patients can go for many months without an AT/AF episode only to experience recurrences which would be difficult to record with intermittent monitoring methods. Despite an improvement in symptoms in all patients following the ablation, many still experienced significant amounts of AT/AF [69]. This long-term trending information enabled the physician to perform additional ablation procedures in patients who did not respond to the initial procedure.

Figure 8.16(a) shows an example of the diagnostic information from implantable devices that can be used to evaluate the efficacy of rhythm control. The total hours of AT/AF per day can be continuously monitored over a period of 14 months before the data gets overwritten. Other diagnostics which can provide additional rhythm control clarity are an AT/AF episode duration histogram, episode start time histogram, and the duration of the longest episode. A key to adequate anticoagulation management may be the early notification of "new onset" AT/AF, i.e., a patient's first episode of AT/AF. With most implantable devices, the clinician does not become aware of the presence of new onset AT/AF until the patient's device is interrogated in the office setting. This could introduce a delay of up to six months from the time of the first AT/AF occurrence to the first opportunity to treat it. Some newer devices have wireless alert capabilities which can automatically send a notification to clinicians when a programmable AT/AF burden threshold is exceeded on a given day. Early notification could allow physicians to intervene more quickly and initiate an anticoagulation regimen which may reduce the risk of stroke in patients who were unaware of their atrial arrhythmias.

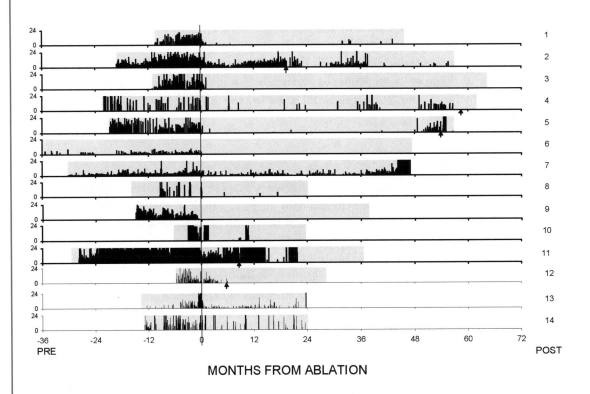

Figure 8.15: Daily AT/AF burden is shown both before and after pulmonary vein ablation in 14 patients. Grey shading indicates the follow-up experience for each patient while the black lines indicate the AT/AF burden experienced each day. Repeat ablation procedures are indicated with arrows.

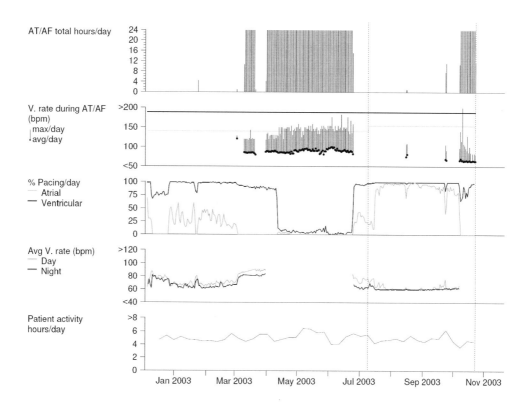

(a)

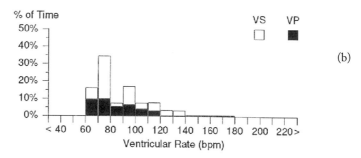

(b)

Figure 8.16: (a) Implantable devices can track measures of rhythm control (AT/AF total hours/day), rate control (ventricular rate during AT/AF), as well as other device and patient parameters over extended periods of time. (b) A histogram of ventricular rates (VS=ventricular sense, VP=ventricular pace) during AT/AF can be used to assess rate control efficacy between follow-up visits.

8.5.3   MONITORING FOR RATE CONTROL

Controlling the rate at which atrial impulses conduct to the ventricle during AT/AF is another valid strategy for treating patients with AT/AF. There are several reasons why it is important to control the ventricular rate. First, rapid and irregular ventricular rates can cause symptoms in many patients. Pharmacologic therapy is generally attempted first to slow and regularize the rate. In cases where drugs cannot adequately control the rate, ablation of the AV node and subsequent pacemaker implantation may be considered. A second reason for maintaining ventricular rate control is to reduce the likelihood of inappropriate shocks being delivered to patients with ICDs. Although it occurs infrequently, rapid and irregular ventricular conduction can occasionally fool the device into thinking that a life-threatening ventricular arrhythmia is present [70]. This could result in a painful and unnecessary shock which could have been avoided with adequate rate control. A third reason is to ensure that patients with CRT devices receive continuous therapy. With CRT, pacing leads in the right and left ventricle are used to synchronize the ventricular contraction and mitigate the effects of heart failure. When rapid intrinsic ventricular rates are caused by conduction of AT/AF, the CRT device may be inhibited from providing its therapy [71].

Implantable devices are capable of continuously monitoring the ventricular rate during AT/AF and reporting this information as a daily tabulation of the average and maximum ventricular rates over the past 14 months (Figure 8.16(a)) or as a histogram of rates since the previous follow-up visit (Figure 8.16(b)). The average ventricular rate during sinus rhythm can also be reported separately for day and night periods (Figure 8.16(a)). Timely notification of poor rate control may help clinicians to prevent the occurrence of symptoms, reduce inappropriate shocks, and ensure continuous CRT therapy in patients with AT/AF. Again, a wireless alert feature can notify the clinician if the average ventricular rate exceeds a specified threshold for a programmable period of time.

8.6   CONCLUSIONS

Implantable devices are capable of continuously sensing electrical activity and utilizing detection algorithms to identify the occurrence of AT/AF. In the future, the performance of the detection algorithms will continue to improve. Consolidating this vast amount of information into a limited number of clinically relevant diagnostics enables clinicians to effectively manage their patients with AT/AF. As the memory storage capacity of implantable devices increases, more relevant information will be stored and the data will be transmitted more efficiently to the medical staff for optimal management. The morbidity associated with implantable devices, especially the "monitoring only" devices, should decrease further by developing devices that are much smaller and easier to implant.

Bibliography

[1] R. G. Hart and J. L. Halperin, "Atrial fibrillation and thromboembolism: A decade of progress in stroke prevention," *Ann. Intern. Med.*, vol. 131, pp. 688–695, 1999.

[2] D. Bialy, M. H. Lehmann, D. N. Schumacher, R. T. Steinman, and M. D. Meissner, "Hospitalization for arrhythmias in the United States: Importance of atrial fibrillation," *J. Am. Coll. Cardiol.*, vol. 19, p. 41A, 1992.

[3] W. B. Kannel, R. D. Abbott, D. D. Savage, and P. M. McNamara, "Coronary heart disease and atrial fibrillation: The Framingham study," *Am. Heart J.*, vol. 106, pp. 389–396, 1983. DOI: 10.1016/0002-8703(83)90208-9

[4] V. Fuster, L. E. Ryden, D. S. Cannom, H. J. Crijns, A. B. Curtis, *et al.*, "ACC/AHA/ESC 2006 guidelines for the management of patients with atrial fibrillation: full text: A report of the American College of Cardiology/American Heart Association Task Force on practice guidelines and the European Society of Cardiology Committee for Practice Guidelines developed in collaboration with the European Heart Rhythm Association and the Heart Rhythm Society," *Europace*, vol. 8, pp. 651–745, 2006.

[5] D. G. Wyse, A. L. Waldo, J. P. DiMarco, M. J. Domanski, Y. Rosenberg, E. B. Schron, J. C. Kellen, H. L. Greene, M. C. Mickel, J. E. Dalquist, and S. D. Corley, "A comparison of rate control and rhythm control in patients with atrial fibrillation," *N. Engl. J. Med.*, vol. 347, pp. 1825–1833, 2002.

[6] S. A. Strickberger, J. Ip, S. Saksena, K. Curry, T. D. Bahnson, and P. D. Ziegler, "Relationship between atrial tachyarrhythmias and symptoms," *Heart Rhythm*, vol. 2, pp. 125–131, 2005. DOI: 10.1016/j.hrthm.2004.10.042

[7] S. A. Rothman, J. C. Laughlin, J. Seltzer, J. S. Walia, R. I. Baman, S. Y. Siouffi, R. M. Sangrigoli, and P. R. Kowey, "The diagnosis of cardiac arrhythmias: A prospective multi-center randomized study comparing mobile cardiac outpatient telemetry versus standard loop event monitoring," *J. Cardiovasc. Electrophysiol.*, vol. 18, pp. 241–247, 2007. DOI: 10.1111/j.1540-8167.2006.00729.x

[8] J. A. Reiffel, R. Schwarzberg, and M. Murry, "Comparison of autotriggered memory loop recorders versus standard loop recorders versus 24-hour Holter monitors for arrhythmia detection," *Am. J. Cardiol.*, vol. 95, pp. 1055–1059, 2005. DOI: 10.1016/j.amjcard.2005.01.025

[9] P. D. Ziegler, J. L. Koehler, and R. Mehra, "Comparison of continuous versus intermittent monitoring of atrial arrhythmias," *Heart Rhythm*, vol. 3, pp. 1445–1452, 2006. DOI: 10.1016/j.hrthm.2006.07.030

[10] C. R. Vasamreddy, D. Dalal, J. Dong, A. Cheng, D. Spragg, S. Z. Lamiy, G. Meininger, C. A. Henrikson, J. E. Marine, R. Berger, and H. Calkins, "Symptomatic and asymptomatic atrial fibrillation in patients undergoing radiofrequency catheter ablation," *J. Cardiovasc. Electrophysiol.*, vol. 17, pp. 134–139, 2006. DOI: 10.1111/j.1540-8167.2006.00359.x

[11] E. J. Benjamin, D. Levy, S. M. Vaziri, R. B. D'Agostino, A. J. Belanger, and P. A. Wolf, "Independent risk factors for atrial fibrillation in a population-based cohort. The Framingham Heart Study," *JAMA*, vol. 271, pp. 840–844, 1994. DOI: 10.1001/jama.271.11.840

[12] G. Senatore, G. Stabile, E. Bertaglia, G. Donnici, A. De Simone, F. Zoppo, P. Turco, P. Pascotto, and M. Fazzari, "Role of transtelephonic electrocardiographic monitoring in detecting short-term arrhythmia recurrences after radiofrequency ablation in patients with atrial fibrillation," *J. Am. Coll. Cardiol.*, vol. 45, pp. 873–876, 2005. DOI: 10.1016/j.jacc.2004.11.050

[13] R. L. Page, T. W. Tilsch, S. J. Connolly, D. J. Schnell, S. R. Marcello, W. E. Wilkinson, and E. L. Pritchett, "Asymptomatic or 'silent' atrial fibrillation: Frequency in untreated patients and patients receiving azimilide," *Circulation*, vol. 107, pp. 1141–1145, 2003. DOI: 10.1161/01.CIR.0000051455.44919.73

[14] F. Roche, J. M. Gaspoz, A. Da Costa, K. Isaaz, D. Duverney, V. Pichot, F. Costes, J. R. Lacour, and J. C. Barthelemy, "Frequent and prolonged asymptomatic episodes of paroxysmal atrial fibrillation revealed by automatic long-term event recorders in patients with a negative 24-hour Holter," *Pacing Clin. Electrophysiol.*, vol. 25, pp. 1587–1593, 2002. DOI: 10.1046/j.1460-9592.2002.01587.x

[15] C. W. Israel, G. Gronefeld, J. R. Ehrlich, Y. G. Li, and S. H. Hohnloser, "Long-term risk of recurrent atrial fibrillation as documented by an implantable monitoring device: Implications for optimal patient care," *J. Am. Coll. Cardiol.*, vol. 43, pp. 47–52, 2004. DOI: 10.1016/j.jacc.2003.08.027

[16] P. Defaye, F. Dournaux, and E. Mouton, "Prevalence of supraventricular arrhythmias from the automated analysis of data stored in the DDD pacemakers of 617 patients: The AIDA study. The AIDA Multicenter Study Group. Automatic Interpretation for Diagnosis Assistance," *Pacing Clin. Electrophysiol.*, vol. 21, pp. 250–255, 1998. DOI: 10.1111/j.1540-8159.1998.tb01098.x

[17] I. C. van Gelder, V. E. Hagens, H. A. Bosker, J. H. Kingma, S. A. Said, J. I. Darmanata, A. J. Timmermans, J. G. Tijssen, and H. J. Crijns, "A comparison of rate control and rhythm control in patients with recurrent persistent atrial fibrillation," *N. Engl. J. Med.*, vol. 347, pp. 1834–1840, 2002. DOI: 10.1056/NEJMoa021375

[18] C. W. Israel, B. Hugl, C. Unterberg, T. Lawo, I. Kennis, D. Hettrick, and S. H. Hohnloser, "Pace-termination and pacing for prevention of atrial tachyarrhythmias: Results from a multi-center study with an implantable device for atrial therapy," *J. Cardiovasc. Electrophysiol.*, vol. 12, pp. 1121–1128, 2001. DOI: 10.1046/j.1540-8167.2001.01121.x

[19] H. Purerfellner, A. M. Gillis, R. Holbrook, and D. A. Hettrick, "Accuracy of atrial tachyarrhythmia detection in implantable devices with arrhythmia therapies," *Pacing Clin. Electrophysiol.*, vol. 27, pp. 983–992, 2004. DOI: 10.1111/j.1540-8159.2004.00569.x

[20] M. A. Wood, P. Moskovljevic, B. S. Stambler, and K. A. Ellenbogen, "Comparison of bipolar atrial electrogram amplitude in sinus rhythm, atrial fibrillation, and atrial flutter," *Pacing Clin. Electrophysiol.*, vol. 19, pp. 150–156, 1996. DOI: 10.1111/j.1540-8159.1996.tb03306.x

[21] K. T. Konings, J. L. Smeets, O. C. Penn, H. J. Wellens, and M. A. Allessie, "Configuration of unipolar atrial electrograms during electrically induced atrial fibrillation in humans," *Circulation*, vol. 95, pp. 1231–1241, 1997.

[22] F. X. Roithinger, A. SippensGroenewegen, M. R. Karch, P. R. Steiner, W. S. Ellis, and M. D. Lesh, "Organized activation during atrial fibrillation in man: Endocardial and electrocardiographic manifestations," *J. Cardiovasc. Electrophysiol.*, vol. 9, pp. 451–461, 1998. DOI: 10.1111/j.1540-8167.1998.tb01836.x

[23] J. M. Baerman, K. M. Ropella, A. V. Sahakian, J. A. Kirsh, and S. Swiryn, "Effect of bipole configuration on atrial electrograms during atrial fibrillation," *Pacing Clin. Electrophysiol.*, vol. 13, pp. 78–87, 1990. DOI: 10.1111/j.1540-8159.1990.tb02006.x

[24] K. M. Ropella, A. V. Sahakian, J. M. Baerman, and S. Swiryn, "The coherence spectrum. A quantitative discriminator of fibrillatory and nonfibrillatory cardiac rhythms," *Circulation*, vol. 80, pp. 112–119, 1989.

[25] K. M. Ropella, A. V. Sahakian, J. M. Baerman, and S. Swiryn, "Effects of procainamide on intra-atrial electrograms during atrial fibrillation: Implications for detection algorithms," *Circulation*, vol. 77, pp. 1047–1054, 1988.

[26] J. Slocum, A. V. Sahakian, and S. Swiryn, "Computer discrimination of atrial fibrillation and regular atrial rhythms from intra-atrial electrograms," *Pacing Clin. Electrophysiol.*, vol. 11, pp. 610–621, 1988. DOI: 10.1111/j.1540-8159.1988.tb04557.x

[27] K. A. Ellenbogen, G. Kay, C.-P. Lau, and B. Wilkoff, *Cardiac Pacing, Defibrillation, and Resynchronization Therapy.* Philadelphia: Saunders/Elsevier, 3rd ed., 2007.

[28] C. R. Kerr and M. A. Mason, "Amplitude of atrial electrical activity during sinus rhythm and during atrial flutter–fibrillation," *Pacing Clin. Electrophysiol.*, vol. 8, pp. 348–355, 1985. DOI: 10.1111/j.1540-8159.1985.tb05769.x

[29] A. L. Beeman, G. Deutsch, and R. F. Rea, "Paradoxical undersensing due to quiet timer blanking," *Heart Rhythm*, vol. 1, pp. 345–347, 2004. DOI: 10.1016/j.hrthm.2004.04.012

[30] R. Willems, P. Holemans, H. Ector, F. Van de Werf, and H. Heidbuchel, "Paradoxical undersensing at a high sensitivity in dual chamber pacemakers," *Pacing Clin. Electrophysiol.*, vol. 24, pp. 308–315, 2001. DOI: 10.1046/j.1460-9592.2001.00308.x

[31] V. Kuhlkamp, V. Dornberger, C. Mewis, R. Suchalla, R. F. Bosch, and L. Seipel, "Clinical experience with the new detection algorithms for atrial fibrillation of a defibrillator with dual chamber sensing and pacing," *J. Cardiovasc. Electrophysiol.*, vol. 10, pp. 905–915, 1999. DOI: 10.1111/j.1540-8167.1999.tb01261.x

[32] M. H. Anderson, F. D. Murgatroyd, K. Hnatkova, B. Xie, S. Jones, E. Rowland, D. E. Ward, A. J. Camm, and M. Malik, "Performance of basic ventricular tachycardia detection algorithms in implantable cardioverter defibrillators: Implications for device programming," *Pacing Clin. Electrophysiol.*, vol. 20, pp. 2975–2983, 1997. DOI: 10.1111/j.1540-8159.1997.tb05469.x

[33] G. Inama, M. Santini, L. Padeletti, G. Boriani, G. Botto, A. Capucci, M. Gulizia, R. Ricci, P. Rizzon, F. Ferri, F. Miraglia, R. Raneri, and A. Grammatico, "Far-field R wave oversensing in dual chamber pacemakers designed for atrial arrhythmia management: Effect of pacing site and lead tip to ring distance," *Pacing Clin. Electrophysiol.*, vol. 27, pp. 1221–1230, 2004. DOI: 10.1111/j.1540-8159.2004.00613.x

[34] C. W. Israel, "Analysis of mode switching algorithms in dual chamber pacemakers," *Pacing Clin. Electrophysiol.*, vol. 25, pp. 380–393, 2002. DOI: 10.1046/j.1460-9592.2002.00380.x

[35] C. T. Lam, C. P. Lau, S. K. Leung, H. F. Tse, and G. Ayers, "Improved efficacy of mode switching during atrial fibrillation using automatic atrial sensitivity adjustment," *Pacing Clin. Electrophysiol.*, vol. 22, pp. 17–25, 1999. DOI: 10.1111/j.1540-8159.1999.tb00295.x

[36] C. P. Lau, S. K. Leung, H. F. Tse, and S. S. Barold, "Automatic mode switching of implantable pacemakers: II. Clinical performance of current algorithms and their programming," *Pacing Clin. Electrophysiol.*, vol. 25, pp. 1094–113, 2002. DOI: 10.1046/j.1460-9592.2002.01094.x

[37] S. K. Leung, C. P. Lau, C. T. Lam, H. F. Tse, M. O. Tang, F. Chung, and G. Ayers, "Programmed atrial sensitivity: A critical determinant in atrial fibrillation detection and optimal automatic mode switching," *Pacing Clin. Electrophysiol.*, vol. 21, pp. 2214–2219, 1998. DOI: 10.1111/j.1540-8159.1998.tb01155.x

[38] R. S. Passman, K. M. Weinberg, M. Freher, P. Denes, A. Schaechter, J. J. Goldberger, and A. H. Kadish, "Accuracy of mode switch algorithms for detection of atrial tachyarrhythmias," *J. Cardiovasc. Electrophysiol.*, vol. 15, pp. 773–777, 2004. DOI: 10.1046/j.1540-8167.2004.03537.x

[39] E. Hammel, C. Hudelo, L. Maillard, B. Besson, and B. Charbonnier, "Appropriate detection of Guidant pacemaker stored electrograms assessed by centralized arrhythmia workstation," *Pacing Clin. Electrophysiol.*, vol. 23, p. 680 (abstract), 2000.

[40] C. W. Israel, D. Gascon, B. Nowak, G. Campanale, P. Pascotto, W. Hartung, and D. Lellouche, "Diagnostic value of stored electrograms in single-lead VDD systems," *Pacing Clin. Electrophysiol.*, vol. 23, pp. 1801–1803, 2000.

[41] D. Bloem and R. Arzbaecher, "Discrimination of atrial arrhythmias using autoregressive modelling," in *Proc. Comput. Cardiol.*, vol. 19, pp. 235–238, IEEE Computer Society, 1992. DOI: 10.1109/CIC.1992.269403

[42] J. Jenkins, K. H. Noh, A. Guezennec, T. Bump, and R. Arzbaecher, "Diagnosis of atrial fibrillation using electrograms from chronic leads: Evaluation of computer algorithms," *Pacing Clin. Electrophysiol.*, vol. 11, pp. 622–631, 1988. DOI: 10.1111/j.1540-8159.1988.tb04558.x

[43] J. Kim, J. Bocek, H. White, B. Crone, C. Alferness, and J. Adams, "An atrial fibrillation detection algorithm for an implantable atrial defibrillator," in *Proc. Comput. Cardiol.*, vol. 22, pp. 169–172, IEEE Computer Society, 1995.

[44] C. D. Swerdlow, W. Schsls, B. Dijkman, W. Jung, N. V. Sheth, W. H. Olson, and B. D. Gunderson, "Detection of atrial fibrillation and flutter by a dual-chamber implantable cardioverter-defibrillator. For the Worldwide Jewel AF Investigators," *Circulation*, vol. 101, pp. 878–885, 2000.

[45] M. M. Morris, B. H. KenKnight, and D. J. Lang, "Detection of atrial arrhythmia for cardiac rhythm management by implantable devices," *J. Electrocardiol.*, vol. 33 (Suppl), pp. 133–139, 2000. DOI: 10.1054/jelc.2000.20305

[46] A. Schuchert, G. Boriani, C. Wollmann, M. Biffi, M. Kuhl, J. Sperzel, S. Stiller, G. Gasparini, and D. Bocker, "Implantable dual-chamber defibrillator for the selective treatment of spontaneous atrial and ventricular arrhythmias: Arrhythmia incidence and device performance," *J. Interv. Card. Electrophysiol.*, vol. 12, pp. 149–156, 2005. DOI: 10.1007/s10840-005-6551-5

[47] K. Kettering, V. Dornberger, R. Lang, R. Vonthein, R. Suchalla, R. F. Bosch, C. Mewis, B. Eigenberger, and V. Kuhlkamp, "Enhanced detection criteria in implantable cardioverter defibrillators: Sensitivity and specificity of the stability algorithm at different heart rates," *Pacing Clin. Electrophysiol.*, vol. 24, pp. 1325–1333, 2001. DOI: 10.1046/j.1460-9592.2001.01325.x

[48] C. D. Swerdlow, M. L. Brown, K. Lurie, J. Zhang, N. M. Wood, W. H. Olson, and J. M. Gillberg, "Discrimination of ventricular tachycardia from supraventricular tachycardia by a downloaded wavelet-transform morphology algorithm: A paradigm for development of implantable cardioverter defibrillator detection algorithms," *J. Cardiovasc. Electrophysiol.*, vol. 13, pp. 432–441, 2002. DOI: 10.1046/j.1540-8167.2002.00432.x

[49] M. R. Gold, S. R. Shorofsky, J. A. Thompson, J. Kim, M. Schwartz, J. Bocek, E. G. Lovett, W. Hsu, M. M. Morris, and D. J. Lang, "Advanced rhythm discrimination for implantable cardioverter defibrillators using electrogram vector timing and correlation," *J. Cardiovasc. Electrophysiol.*, vol. 13, pp. 1092–1097, 2002. DOI: 10.1046/j.1540-8167.2002.01092.x

[50] G. J. Klein, J. M. Gillberg, A. Tang, S. Inbar, A. Sharma, C. Unterberg-Buchwald, P. Dorian, H. Moore, F. Duru, E. Rooney, D. Becker, K. Schaaf, and D. Benditt, "Improving SVT discrimination in single-chamber ICDs: A new electrogram morphology-based algorithm," *J. Cardiovasc. Electrophysiol.*, vol. 17, pp. 1310–1319, 2006. DOI: 10.1111/j.1540-8167.2006.00643.x

[51] M. A. Lee, R. Corbisiero, D. R. Nabert, J. A. Coman, M. C. Giudici, G. F. Tomassoni, K. T. Turk, D. J. Breiter, and Y. Zhang, "Clinical results of an advanced SVT detection enhancement algorithm," *Pacing Clin. Electrophysiol.*, vol. 28, pp. 1032–1040, 2005. DOI: 10.1111/j.1540-8159.2005.00219.x

[52] E. Aliot, R. Nitzsche, and A. Ripart, "Arrhythmia detection by dual-chamber implantable cardioverter defibrillators. A review of current algorithms," *Europace*, vol. 6, pp. 273–286, 2004. DOI: 10.1016/j.eupc.2004.02.005

[53] C. D. Swerdlow, "Supraventricular tachycardia-ventricular tachycardia discrimination algorithms in implantable cardioverter defibrillators: State-of-the-art review," *J. Cardiovasc. Electrophysiol.*, vol. 12, pp. 606–612, 2001. DOI: 10.1046/j.1540-8167.2001.00606.x

[54] P. A. Friedman, R. L. McClelland, W. R. Bamlet, H. Acosta, D. Kessler, T. M. Munger, N. G. Kavesh, M. Wood, E. Daoud, A. Massumi, C. Schuger, S. Shorofsky, B. Wilkoff, and M. Glikson, "Dual-chamber versus single-chamber detection enhancements for implantable defibrillator rhythm diagnosis: The detect supraventricular tachycardia study," *Circulation*, vol. 113, pp. 2871–2879, 2006. DOI: 10.1161/CIRCULATIONAHA.105.594531

[55] D. Duverney, J. M. Gaspoz, V. Pichot, F. Roche, R. Brion, A. Antoniadis, and J. C. Barthelemy, "High accuracy of automatic detection of atrial fibrillation using wavelet transform of heart rate intervals," *Pacing Clin. Electrophysiol.*, vol. 25, pp. 457–462, 2002. DOI: 10.1046/j.1460-9592.2002.00457.x

[56] J. Hayano, F. Yamasaki, S. Sakata, A. Okada, S. Mukai, and T. Fujinami, "Spectral characteristics of ventricular response to atrial fibrillation," *Am. J. Physiol.*, vol. 273, pp. H2811–2816, 1997.

[57] K. Tateno and L. Glass, "Automatic detection of atrial fibrillation using the coefficient of variation and density histograms of RR and deltaRR intervals," *Med. Biol. Eng. & Comput.*, vol. 39, pp. 664–671, 2001. DOI: 10.1007/BF02345439

[58] T. Anan, K. Sunagawa, H. Araki, and M. Nakamura, "Arrhythmia analysis by successive RR plotting," *J. Electrocardiol.*, vol. 23, pp. 243–248, 1990. DOI: 10.1016/0022-0736(90)90163-V

[59] A. S. Chishaki, K. Sunagawa, K. Hayashida, M. Sugimachi, and M. Nakamura, "Identification of the rate-dependent functional refractory period of the atrioventricular node in simulated atrial fibrillation," *Am. Heart J.*, vol. 121, pp. 820–826, 1991. DOI: 10.1016/0002-8703(91)90194-M

[60] J. Hayano, S. Sakata, A. Okada, S. Mukai, and T. Fujinami, "Circadian rhythms of atrioventricular conduction properties in chronic atrial fibrillation with and without heart failure," *J. Am. Coll. Cardiol.*, vol. 31, pp. 158–166, 1998. DOI: 10.1016/S0735-1097(97)00429-4

[61] K. Seidl, M. Rameken, S. Breunung, J. Senges, W. Jung, D. Andresen, A. van Toor, A. D. Krahn, and G. J. Klein, "Diagnostic assessment of recurrent unexplained syncope with a new subcutaneously implantable loop recorder. Reveal-investigators," *Europace*, vol. 2, pp. 256–262, 2000. DOI: 10.1053/eupc.2000.0108

[62] B. K. Bootsma, A. J. Hoelsen, J. Strackee, and F. L. Meijler, "Analysis of R-R intervals in patients with atrial fibrillation at rest and during exercise," *Circulation*, vol. 41, pp. 783–794, 1970.

[63] F. L. Meijler, J. Jalife, J. Beaumont, and D. Vaidya, "AV nodal function during atrial fibrillation: The role of electrotonic modulation of propagation," *J. Cardiovasc. Electrophysiol.*, vol. 7, pp. 843–861, 1996. DOI: 10.1111/j.1540-8167.1996.tb00597.x

[64] L. Toivonen, A. Kadish, W. Kou, and F. Morady, "Determinants of the ventricular rate during atrial fibrillation," *J. Am. Coll. Cardiol.*, vol. 16, pp. 1194–1200, 1990.

[65] K. M. Stein, J. Walden, N. Lippman, and B. B. Lerman, "Ventricular response in atrial fibrillation: Random or deterministic?," *Am. J. Physiol.*, vol. 277, pp. H452–458, 1999.

[66] M. Duytschaever, C. Dierickx, and R. Tavernier, "Variable atrioventricular block during atrial flutter: What is the mechanism?," *J. Cardiovasc. Electrophysiol.*, vol. 13, pp. 950–951, 2002. DOI: 10.1046/j.1540-8167.2002.00950.x

[67] S. Sarkar, D. Ritscher, and R. Mehra, "A detector for a chronic implantable atrial tachyarrhythmia monitor," *IEEE Trans. Biomed. Eng.*, vol. 55, 2008 (in press). DOI: 10.1109/TBME.2007.903707

[68] A. L. Goldberger, L. A. Amaral, L. Glass, J. M. Hausdorff, P. C. Ivanov, R. G. Mark, J. E. Mietus, G. B. Moody, C. K. Peng, and H. E. Stanley, "PhysioBank, PhysioToolkit, and PhysioNet: Components of a new research resource for complex physiologic signals," *Circulation*, vol. 101, pp. E215–220, 2000.

[69] H. Purerfellner, J. Aichinger, M. Martinek, H. J. Nesser, P. Ziegler, J. Koehler, E. Warman, and D. Hettrick, "Quantification of atrial tachyarrhythmia burden with an implantable pacemaker before and after pulmonary vein isolation," *Pacing Clin. Electrophysiol.*, vol. 27, pp. 1277–1283, 2004. DOI: 10.1111/j.1540-8159.2004.00620.x

[70] R. Willems, M. L. Morck, D. V. Exner, S. M. Rose, and A. M. Gillis, "Ventricular high-rate episodes in pacemaker diagnostics identify a high-risk subgroup of patients with tachy-brady syndrome," *Heart Rhythm*, vol. 1, pp. 414–421, 2004. DOI: 10.1016/j.hrthm.2004.06.004

[71] B. P. Knight, A. Desai, J. Coman, M. Faddis, and P. Yong, "Long-term retention of cardiac resynchronization therapy," *J. Am. Coll. Cardiol.*, vol. 44, pp. 72–77, 2004. DOI: 10.1016/j.jacc.2004.03.054

Abbreviations

A/D	analog-to-digital (conversion)
ABS	average beat subtraction
ACG	atriocardiogram
AF	atrial fibrillation
ANOVA	analysis of variance
ANS	autonomic nervous system
APD	action potential duration
ApEn	approximate entropy
ARC	atrial rhythm classification
ASC	activation space constant
AT	atrial tachycardia
ATP	antitachycardia pacing
A:V	atrial to ventricular
AV	atrioventricular
BEM	boundary element method
bpm	beats per minute
BSPM	body surface potential mapping
CC	crosscorrelation
CCE	corrected conditional entropy
CD	correlation dimension
CE	conditional entropy
CHD	coronary heart disease
CHF	coronary heart failure
CRN	Courtemanche, Ramirez, Nattel (model)
CRT	cardiac resynchronization therapy
CS	coronary sinus
CV	conduction velocity
CVD	Choie–Williams distribution
DF	dominant frequency
ECG	electrocardiogram
EDL	equivalent double layer
EGM	electrogram
EMI	electromagnetic interference
ERP	effective refractory period

FFT	fast Fourier transform
FRP	functional refractory period
HMM	hidden Markov model
HRSH	heart rate stratified histogram
HRV	heart rate variability
ICA	independent component analysis
ICD	implantable cardioverter defibrillator
ILR	implanted loop recorder
LA	left atrium
LAW	local activation wave
LE	lower envelope
LP	level of predictability
LVH	left ventricular hypertrophy
MCE	mutual-conditional entropy
MR	magnetic resonance
MSC	magnitude-squared coherence
MSE	mean-square error
NLA	nonlinear association
NN	normal-to-normal
NSR	normal sinus rhythm
OACG	optimal atriocardiogram
PCA	principal component analysis
PDC	peak dominant change
PG	peak gap
pNN50	percentage of normal-to-normal RR intervals greater than 50 ms.
PV	pulmonary vein
PVR	peak value ratio
PWD	P wave duration
RA	right atrium
RMS	root mean-square
rMSSD	root mean-square difference of successive normal-to-normal RR intervals
SampEn	sample entropy
SAN	sinoatrial node
SDNN	standard deviation of normal-to-normal RR intervals
SE	Shannon entropy
SNR	signal-to-noise ratio
SOBI	second-order blind identification
SPA	second peak amplitude
SPP	second peak position

SR	sinus rhythm
STFT	short-term Fourier transform
SVD	singular value decomposition
SVT	supraventricular tachycardia
TMP	transmembrane potential
XWVD	cross Wigner–Ville distribution
VCG	vectorcardiogram
VT	ventricular tachycardia
WCT	Wilson central terminal
WVD	Wigner–Ville distribution

Index

Printed in the United States
by Baker & Taylor Publisher Services